SpringerBriefs in Applied Sciences and Technology

T0225754

SpringerBriefs present concise summaries of cutting-edge research and practical applications across a wide spectrum of fields. Featuring compact volumes of 50 to 125 pages, the series covers a range of content from professional to academic.

Typical publications can be:

- A timely report of state-of-the art methods
- An introduction to or a manual for the application of mathematical or computer techniques
- A bridge between new research results, as published in journal articles
- A snapshot of a hot or emerging topic
- An in-depth case study
- A presentation of core concepts that students must understand in order to make independent contributions

SpringerBriefs are characterized by fast, global electronic dissemination, standard publishing contracts, standardized manuscript preparation and formatting guidelines, and expedited production schedules.

On the one hand, **SpringerBriefs in Applied Sciences and Technology** are devoted to the publication of fundamentals and applications within the different classical engineering disciplines as well as in interdisciplinary fields that recently emerged between these areas. On the other hand, as the boundary separating fundamental research and applied technology is more and more dissolving, this series is particularly open to trans-disciplinary topics between fundamental science and engineering.

Indexed by EI-Compendex, SCOPUS and Springerlink.

More information about this series at http://www.springer.com/series/8884

Roya Ahmadiahangar · Argo Rosin ·
Ivo Palu · Aydin Azizi

Demand-side Flexibility in Smart Grid

 Springer

Roya Ahmadiahangar
Department of Electrical Power Engineering
and Mechatronics
Tallinn University of Technology
Tallinn, Estonia

Ivo Palu
Department of Electrical Power Engineering
and Mechatronics
Tallinn University of Technology
Tallinn, Estonia

Argo Rosin
Department of Electrical Power Engineering
and Mechatronics
Tallinn University of Technology
Tallinn, Estonia

Aydin Azizi
School of Engineering, Computing
and Mathematics
Oxford Brookes University
Oxford, UK

ISSN 2191-530X   ISSN 2191-5318 (electronic)
SpringerBriefs in Applied Sciences and Technology
ISBN 978-981-15-4626-6   ISBN 978-981-15-4627-3 (eBook)
https://doi.org/10.1007/978-981-15-4627-3

This Springer imprint is published by the registered company Springer Nature Singapore Pte Ltd.
The registered company address is: 152 Beach Road, #21-01/04 Gateway East, Singapore 189721,
Singapore

Contents

Chapter 1
Challenges of Smart Grids Implementation

1.1 Overview

Increasing share of renewable energy resources, implementation of new technologies and data management methods in power system, development of communication systems from one side, and higher demand of electricity and concerns for increasing existing transmission lines while maintaining grid stability and reliability have been the main motivations for moving toward smart grid.

This chapter discusses the main definitions, components, advantages, and challenges of implementing smart grids.

1.2 Definitions

There is no single definition for smart grid in the litreture. The European technology platform describes Smart Grid as [1], "A Smart Grid is an electricity network that can intelligently integrate the actions of all users connected to it—generators, consumers, and those that do both—in order to efficiently deliver sustainable, economic, and secure electricity supplies."

Another definition of Smart Grid is "A Smart Grid uses sensing, embedded processing, and digital communications to enable the electricity grid to be observable (able to be measured and visualized), controllable (able to manipulated and optimized), automated (able to adapt and self heal), fully integrated (fully interoperable with existing systems, and with the capacity to incorporate a diverse set of energy sources) [1]."

Regarding the U.S. department of energy is, "A Smart Grid uses digital technology to improve reliability, security, and efficiency (both economic and energy) of the

© The Author(s), under exclusive license to Springer Nature Singapore Pte Ltd. 2020
R. Ahmadiahangar et al. *Demand-side Flexibility in Smart Grid*,
SpringerBriefs in Applied Sciences and Technology,
https://doi.org/10.1007/978-981-15-4627-3_1

electrical system from large generation, through the delivery systems to electricity consumers and a growing number of distributed generation and storage resources [1]."

IECs definition for Smart Grid is, "The Smart Grid is a developing network of transmission lines, equipment, controls, and new technologies working together to respond immediately to our twenty-first century demand for electricity [1]."

IEEE definition for Smart Grid is, "The smart grid is a revolutionary undertaking entailing new communications and control capabilities, energy source, generation models, and adherence to cross-jurisdictional regulatory structures [1]."

The main charecteristics of Smart Grid include the following:

- Integration of Renewable Energy Sources (RES) as the main source of energy
- Enabling Demand-side Management (DSM) and Demand Response (DR) with providing information for customers through communication infrastructure and bidirectional power flow
- Resiliency during cyber or physical attacks, disasters and delivers energy to consumers with enhanced levels of security and reliability

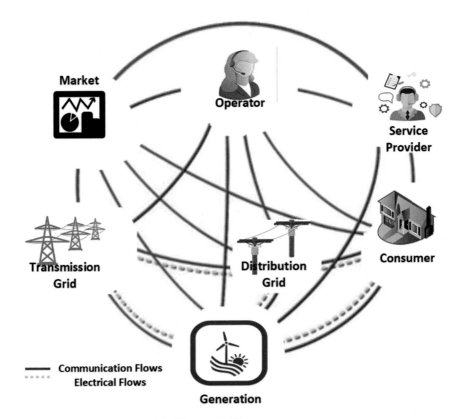

Fig. 1.1 NIST conceptual model of Smart grid [2]

- Increasing efficiency and decreasing loss and maintenance costs.

Interaction of roles in different Smart Grid domains through secure communication are presented in the NIST conceptual model of the Smart grid (Fig. 1.1) [2]. As shown in Fig. 1.1, there are seven main domains, including operations, service providers, customer, distribution, generation, transmission, and markets. Electrical and communication flows are also indicated in Fig. 1.1. It must be noted that the main feature of the NIST conceptual model is the presence of communication infrastructure between all domains. The Smart Grid is actually an electric system that uses information, two-way, cyber-secure communication technologies, and computational intelligence in an integrated fashion across the entire spectrum of the energy system from the generation to the ends of consumption [3].

IEEE Smart Grid also presented the IEEE Smart Grid Domains created by IEEE Smart Grid members as shown in Fig. 1.2 [4]. Based on [4], eight different domains are presented: Operations, Markets, Transmission, Bulk Generation, Non-Bulk Generation, Distribution, Customer, Service Provider, and Foundational Support Systems. The main differences between the NIST model and the IEEE model can be

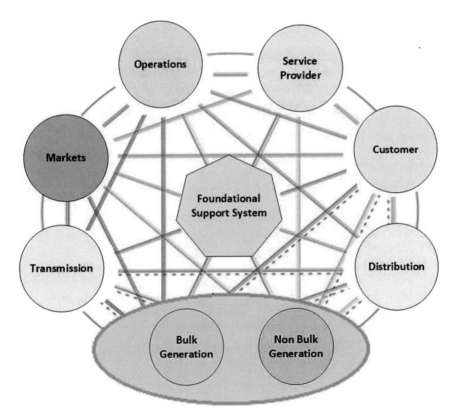

Fig. 1.2 IEEE Smart Grid domains [4]

counted as, in the IEEE model, there is a foundational support systems domain, which is connected to all other domains with two-way communication system. This domain consists of several sub-domains including the following:

- Architecture (Interoperability, Usability, etc.)
- Business Process Re-engineering
- Communication Systems
- Control Systems
- Economic Justification, Cost Recovery Models
- Education and Training
- Environmental Impacts and Efficiency
- Information and Data Management
- Strategy, Policy, Procedure, and Standards
- System Resiliency (Cyber Security, Critical Infrastructure Protection, and Reliability Compliance).

1.3   Micro Grid [5]

The microgrid is usually defined as a small network of loads and distributed energy resources (DER), connected to the main grid but with the ability to operate reliably independently [6]. Main advantages of microgrids are higher supply reliability for consumers, resiliency, and power quality and lower costs and environmental emissions [7].

Microgrids are also becoming increasingly common in universities. Figure 1.3 shows the Tallinn University of Technology's Microgrid configuration. As shown in Fig. 1.3, this microgrid consists of a battery energy storage system, hydrogen energy storage system, fuelcell, flywheel, wind and PV generation units and controllable loads [8, 9].

The main concerns of the control and management of microgrids include energy management, load forecasting [10] stability [11], power quality, power flow control [12], islanding detection, synchronization, and system recovery [13]. The potential complexity of such a system due to possible interactions between intelligent equipment and the power grid, high penetration of DER [14], demand-side management, and market operation requires precise modeling and analysis before practical implementation [15, 16]. As an example, the behavior of the system when disconnected from the power grid must be determined. Frequency control in disconnected or faulty modes is also a main subject of research.

The main advantages of microgrids are as follows:

- Microgrids can provide electric service to regions and communities that are currently unserved
- The use of both electricity and heat permitted by the close proximity of the generator to the user can increase the overall energy efficiency

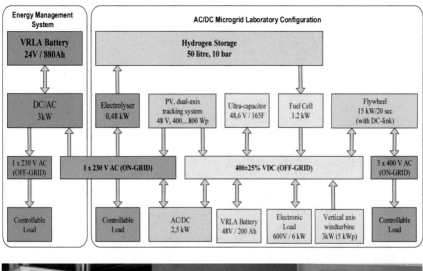

Fig. 1.3 Tallinn University of Technology's Microgrid configuration [5]

- Can provide substantial savings and cut emissions
- Microgrid can facilitate the use of renewable energy sources
- Power generation units are small and are located in close proximity to load
- Can provide high quality and reliable energy supply to critical loads
- Large land use impacts are avoided
- Large transmission build-out may be reduced and transmission losses can be reduced

- Microgrid meets major grid creating a firm electricity supply and subordinate load-balancing area with a single interconnect
- Bidirectional electricity flow allowing ancillary services on either absorption or voltage support
- Creates a power product rather than single plant output
- Can add power quality to both grid and "behind the fence" loads.

The main disadvantage of typical analyzing tools of microgrids (software simulations, prototypes, and pilot projects) is the limited ability to test all interconnection issues. In this context, Real-Time (RT) simulations and Hardware-in-the-loop (HIL) technology are beneficial mainly because of their easily reconfigurable test environment. In latter, all system variables are accessible and there is a good possibility of testing different scenarios, and cases with the same hardware setup [17, 18]. It is also worth mentioning that an RT simulation is a promising approach for validating advance and complex control strategies designed for microgrid and also determining exact values of control parameters and debugging them. As a result, significant cost reduction can be achieved by eliminating control system errors.

In recent years, with the electric power system development toward the smart grid, there is an increasing interest in microgrids as they are considered flexible, intelligent, active participants, secure, environmentally friendly, and economical. The main components of microgrids are distributed generation units (mostly renewable energy sources and energy storage systems [19–21]), flexible and inflexible loads, control and communication systems, and protection units. The microgrids operation can be divided into two general modes, namely grid-connected and islanded mode [6, 22]. The main difference between these operation modes is that in the grid-connected mode the purpose is to compensate renewable energy source fluctuations through storage units [23] and Demand Response (DR) programs in a cost-optimized, environment friendly, and customer satisfying solution, and generally acting like the controllable load from the grid point of view. On the other hand, in islanded mode, the main purpose is maintaining the power balance between generation and demand without grid support.

Some references go through proposing optimization methods for both operation modes [24–27]. A study in [28] presents a business model for the optimal operation of multi-party microgrids and different structures and unique operating goals. In [29], reliability assessment of microgrids is evaluated and improved in both operating modes.

While maintaining the stability of microgrids is important in operation modes [30, 31], all stability parameters like voltage and frequency must be controlled by microgrid independent from the main grid in islanded mode [32, 33] Conventional droop control used for power-sharing among DER may cause undesirable voltage and frequency deviations; the study in [34] presents multiple drooping strategies and their advantages. The power-sharing between diverged DER becomes extremely critical particularly when the response time of the micro-sources differs significantly, in this regard [35] presents a control technique for proportional load sharing in the islanded-mode operation of the microgrid. A decentralized sliding mode control of islanded AC microgrids affected by unknown load dynamics and model uncertainties is presented in [36]. Another solution for maintaining grid stability in islanded mode is to implement cooperative control [37] and provide reference value to a certain number of DGs or active loads in a microgrid, so-called pinning control [38, 39]. FACTS are also used in different studies for improving load-balancing capability of an islanded microgrid. The study in [40] proposed an optimized dynamic PI-controller based on fuzzy logic and seeker optimization approach in a D-STATCOM integrated islanded microgrid to implement economic load sharing.

The most critical operating point of microgrids is transitioned from grid-connected to islanded mode and vice a versa [41]. The study in [42] presents the design, operational and control requirements, and management strategies required for seamless transition mode in commercial microgrids. In this regard, the transition mode can be added to two previously defined operation modes. One issue in this area is islanding detection. Possessing a no detection zone by many of islanding detection studies is proposed in [43] and as a solution, a new technique capable of detecting islanding events in less than 7.5 cycles with an accuracy of 100% justifying its real-time application is presented. The study in [42] concluded the necessity of different control and management strategies for operation modes. The study in [44] considers the grid-connected and islanded modes of voltage-controlled voltage source inverters (VC-VSIs) and current-controlled voltage source inverters (CC-VSIs) connected to a microgrid and proposed a control method for a seamless transition in a single-phase microgrid. In [45], a hierarchical control scheme is proposed for robust operation of microgrid and a seamless transition between operation modes. Many types of energy storage devices having large power density can compensate transient power. The studies in [46, 47] considered battery energy storage, compressed air storage systems, flywheel, and super capacitor-based storage units to address both steady and transient power demands.

1.4 Advantages of Smart Grid

The main advantages of smart grid can be divided into reducing environmental impacts, improving reliability, and increasing efficiency. Following sections describe these advantages.

1.4.1 Reducing Environmental Impacts

Several environmental impact assessments have been made to support policy on the deployment of the smart grid [48] Reducing CO_2 and environmental effects of the power system in one of the most important drivers of the smart grid. In this regard, the environmental benefits of smart grid are as follows:

- **Integrating renewable energy sources** in power systems and increasing their share of generation will result in a significant decrease in CO_2 and other emissions. Integrating RES especially in bulk amounts requires real-time control management of the grid that is possible in the context of smart grid. From the environmental impacts point of view, with increasing share of RES in generation, there would be a decrease in the share of conventional power plants, which are mostly fossil-fuel based, thus, this results in lower producing lower emissions electricity generation [49].
- **Decreasing needs for new transmission lines and power plants**

The Smart grid will decrease the needs for new transmission lines and power plants for different reasons. One is that with development in the use of distributed generations, the generation part would be more distributed and local, thus the need for constructing new transmission lines and power plants decrease. The other reason is that in the smart grid vision, there would be the possibility of full controllability and observability, which results in higher efficiency and lower developing needs.

- **Possibility of using PHEVs and V2G**

PHEV and V2g are smart grid technologies that can provide different services between transport and electrical grid infrastructures [50]. These services are beneficial for both grid utilities and electric vehicle owners.

- **Decreasing Demand Peaks**

Smart grid enables demand response programs by giving real-time market signal access to customers and the opportunity to shift their consumption to a lower price time. Reducing peak load has different benefits for stockholders. For consumer side, it decreases their electricity cost, for generating companies it will result in lower loss, and finally, for utilities, lower investment costs in expanding transmission and distribution infrastructure [51–54].

1.4.2 Improving Reliability

One of the main advantages of integrating DER is increasing power adequacy and improving generation reliability. On the grid side, the presence of communication infrastructure enables fast control of the grid in presence of contingencies and congestion prevention. The report in [55] claims through a survey of 200S G projects, 70% of pilots have experienced enhanced reliability up to 9%.

1.4.3 Increasing Efficiency

Smart grid reduces inefficiencies in energy delivery in both transmission and distribution systems by decreasing loss and congestions. Modern control systems and increasing share of RES will result in increasing efficiency of generation. Finally, there is also a great opportunity to increase demand-side efficiency with different demand-side management programs.

1.5 Main Challenges of Smart Grids Implementations

High investment costs, maintaining security and privacy, and operational complexity can be considered as the main challenges of future smart grids.

1.5.1 Investment Costs

Smart grids consist of wide range of technologies in all domains, from consumers to transmission and distribution grids; therefore it needs investment in different sections including the following:

- **Communication infrastructure and smart meters**: The smart grid, being a vast system, may utilize various communications and networking technologies with its applications, which include both wired and wireless communications. Different communication technologies have different pros and cons, GSM and GPRS have coverage range of up to 10 km but they lack in data rates. 3G requires a costlier spectrum, whereas ZigBee is limited by coverage range of 30–50 m only and is usually used for home energy management purposes. Power line communication and other wired types, don't have typical issues of wireless communication but they have interferences issue. Despite being fast and secure, optical fibre is expensive. And Router-based RF technology is still in early stage of development [56].

  In addition, smart meter deployment is the first step to enable the demand response program and it needs high initial investment cost. Changing all regular meters of customers to smart meters and building several data centers to integrate all the data in control centers.

- **Grid Protection**: Traditionally, the power flow is in one direction in power systems. In the smart grid, bidirectional power flow is a result of the presence of distributed generation in the demand side and the possibility of selling demand generation to the distribution utility and in a higher level sending the excess power of distribution grid to the transmission grid. This will totally change the protection philosophy. So changing many protection devices and implementing new protection coordination is necessary.

- **Storage systems**: Due to non-uniformity of RES generation, there would be a demand for different storage technologies in smart grid. Most common storage technologies are batteries, flywheels, thermal storage, and hydrogen storage. Increasing storage capacity in the grid requires some major investments.

1.5.2 Security and Privacy

Connecting electricity and cyber networks together, will bring different challenges. Most important ones are availability, integrity and confidentiality [56]. Availability refers to reliable and timely access to database and other information; integrity includes protection from improper format/modification/destruction of information; confidentiality refers to security of information from unauthorized access. Since in smart grid vision, all nodes are supposed to be connected to each other , this will bring cybersecurity concerns.

Well-known cyber threats are hackers, zero day, malware, etc.

Meanwhile, in the consumer side, increasing instalment of smart meters create huge amount of data, which being shared between different entities. Leakage of this data may leads to several serious problems [57].

1.5.3 Data Management

In the vision of Smart grids utilities face a major challenge on how to deal with data management, since they generate a large quantity of data, for example, SCADA system collects data every 2–5 s, AMI system collects data every 1–15 min, etc. [58].

- **Standards and interoperability**: Different protocols with different definitions and communication techniques are presented in the smart grid. Several attempts have been made to standardize these; IEC 61850, IEC 61970/61968, IEEE 1815 and IEEE 2030.5 are the most important ones.
- **Massive data**: Processing and storing the huge amount of data in smart grid is an important issue. Another issue related to the big data is finding a fusion method for multisource dataset, which has different modalities, formats, and representations [57]. Another issue is that applications, like transient stability, the reaction time is in the order of milliseconds; in this case, latency resulting from massive amount of data and several optimization and control actions must be solved. New data analysis skills are needed in the utilities in order to make full use of the data [59].

1.5.4 Stability and Decreasing Rate of Flexibility

Maintaining grid stability is the main operational challenge of smart grid. Traditionally synchronous generators maintain power system stability. Synchronous generators support the grid with their inertia, droop control and reactive power compensation. With an increasing share of RES in generation, which results in decreasing share of synchronous machines, new stability challenges impose to the grid:

- **Power balance**: Traditionally, generation units are supposed to follow the demand in order to maintain the power balance. This is probably one of the most complicated problems of the power system since it depends on several factors like load forecasts, unit commitments, transmission contingencies and constraints, etc. To make it even worse, in smart grid, generation units (mostly RES) are not always available and generation fluctuation would be even more than load. As a result, more flexibility is needed in the power system to overcome all these unbalances.
- **Voltage stability**: Frequent and intermittent fluctuations of active power generation from RES results in major voltage deviation in distribution level [60]. Traditional voltage regulators in the grid-like tap changers and capacitor banks are not capable of compensating fast variations coming from RES [61]. Using produced reactive power of RES power electronic interface to overcome voltage instabilities is ongoing research.
- **Frequency stability**: One of the main drawbacks of integrating large amounts of RES to the grid is that they are connected to the power system through power electronic devices, thus they cannot support power system inertia [62]. It must be noted that low system inertia is related to a faster rate of change of frequency and larger frequency deviations. In addition, conventional frequency method, droop control, is not accessible with RES. Thus, maintaining grid frequency stability requires new solutions (virtual inertia as an example) and control methods.

1.6 Organization of the Book

After the brief introduction, the rest of the book has been organized in the following manner. Chapter 2 describes the concept of flexibility in the power systems; Chap. 3 presents modeling and quantifying methods for power system flexibility. Chapter 4 presents different forecasting available demand-side flexibility methods, and finally, Chap. 5 presents possible solutions for increasing demand-side flexibility.

Acknowledgements This work has been supported by the European Commission through the H2020 project Finest Twins (grant No. 856602).

References

1. G. Dileep, A survey on smart grid technologies and applications. Renew. Energy **146**, 2589–2625 (2020)
2. Office of the National Coordinator for Smart Grid Interoperability Engineering Laboratory in collaboration with Physical Measurement Laboratory and Information Technology Laboratory. NIST Framework and Roadmap for Smart Grid Interoperability Standards, Release 2.0 (NIST Special Publication, 1108R2, 2012)
3. https://www.nist.gov/programs-projects/smart-grid-communications-0. NIST, 7 May 2018. [Online]. Accessed 6 Sept 2019
4. https://smartgrid.ieee.org/domains. [Online]. Accessed 10 June 2019
5. R. AhmadiAhangar, A. Rosin, A.N. Niaki, I. Palu, T. Korõtko, A review on real-time simulation and analysis methods of microgrids. Int. Trans. Electr. Energy Syst. **29**(11), e12106 (2019)
6. Summary Report: 2012 DOE Microgrid. Available: http://energy.gov/sites/prod/files/2012%20Microgrid%20workshop%20Report%2009102012.pdf. Accessed 13 Nov 2014. (2012)
7. S. Parhizi, H. Lotfi, A. Khodaei, S. Bahramirad, State of the art in research on microgrids: a review. IEEE Access **3**, 890–925 (2015)
8. D. Lebedev, A. Rosin, L. Kütt, Simulation of real time electricity price based energy management system, in *IECON 2016—42nd Annual Conference of the IEEE Industrial Electronics Society*, Florence, Italy (2016)
9. D. Lebedev, A. Rosin, Practical use of the energy management system with day-ahead electricity prices, in *IEEE 5th International Conference on Power Engineering, Energy and Electrical Drives (POWERENG)*, Riga, Latvia (2015)
10. M. Mohammadi, F. Talebpour, E. Safaee, N. Ghadimi, O. Abedinia, Small-scale building load forecast based on hybrid forecast engine. Neural Process. Lett. **48**(1), 329–351 (2017)
11. O. Abedinia, M. Bekravi, N. Ghadimi, Intelligent controller based wide-area control in power system. Int. J. Uncertainty, Fuzziness Knowl. Based Syst. **25**(1), 1–30 (2017)
12. R. Ahmadi, A. Sheykholeslami, A.N. Niaki, H. Ghaffari, Power flow control and solutions with dynamic flow controller, in *Electric Power Conference, EPEC 2008*. IEEE Canada, Vancouver (2008)
13. T. Logenthiran, R.T. Naayagi, W.L. Woo, V.T. Phan, K. Abidi, Intelligent control system for microgrids using multiagent system. IEEE J. Emerg. Sel. Top. Power Electron. **3**(4), 1036–1045 (2015)
14. E. Imaie, A. Sheikholeslami, R. Ahmadi Ahangar, Improving short-term wind power prediction with neural network and ICA algorithm and input feature selection. J. Adv. Comput. Res. **5**(3), 13–34 (2014)
15. J. Xiao, P. Wang, L. Setyawan, X. Qianwen, Multi-level energy management system for real-time scheduling of dc microgrids with multiple slack terminals. IEEE Trans. Energy Convers. **31**(1), 392–401 (2016)
16. M.H. Cintuglu, T. Youssef, Development and application of a real-time testbed for multiagent system interoperability: a case study on hierarchical microgrid control. IEEE Trans. Smart Grid **9**(3), 1759–1765 (2018)
17. M.C. Magro, M. Giannettoni, P. Pinceti, Real time simulator for microgrids. Electric Power Syst. Res. **160**, 381–396 (2018)
18. F. Huerta, R.L. Tello, M. Prodanovic, Real-time power-hardware-in-the-loop implementation of variable-speed wind turbines. IEEE Trans. Industr. Electron. **64**(3), 1893–1904 (2017)
19. A. Rahmoun, A. Armstorfer, H. Biechi, A. Rosin, Mathematical modeling of a battery energy storage system in grid forming mode, in *IEEE 58th International Scientific Conference on Power and Electrical Engineering of Riga Technical University (RTUCON)* (2017)
20. I. Roasto, T. Lehtla, T. Moller, A. Rosin, Control of ultracapacitors energy exchange, in *12th International Power Electronics and Motion Control Conference, EPE-PEMC 2006*

21. S. Choudhury, T.P. Dash, P. Bhowmik, P.K. Rout, A novel control approach based on hybrid fuzzy logic and seeker optimization for optimal energy management between micro-sources and supercapacitor in an islanded Microgrid. J. King Saud Univ. Eng. Sci. https://doi.org/10.1016/j.jksues.2018.03.006 (in Press, 2019)

22. Q. Jiang, M. Xue, G. Geng, Energy management of microgrid in grid-connected and stand-alone modes. IEEE Trans. Power Syst. **28**(3), 3380–3389 (2013)

23. R. Ahmadi, F. Ghardashi, D. Kabiri, A. Sheykholeslami, H. Haeri, Voltage and frequency control in smart distribution systems in presence of DER using flywheel energy storage system, in *IET Digital Library*, pp. 1307–1307 (2013)

24. A. Selakov, D. Bekut, A.T. Sarić, A novel agent-based microgrid optimal control for grid-connected, planned island and emergency island operations. Int. Trans. Electr. Energy Syst. **26**, 1999–2022 (2016)

25. F. Zhang, H. Zhao, M. Hong, Operation of networked microgrids in a distribution system. CSEE J. Power Energy Syst. **1**(4), 12–21 (2015)

26. S. Chandak, P. Bhowmik, M. Mishra, P.K. Rout, Autonomous microgrid operation subsequent to an anti-islanding scheme. Sustain. Cities Soc. **39**, 430–448 (2018)

27. K. Peterson, R. Ahmadiahangar, N. Shabbir, T. Vinnal, Analysis of microgrid configuration effects on energy efficiency, in *2019 IEEE 60th International Scientific Conference on Power and Electrical Engineering of Riga Technical University (RTUCON)*, Riga (2019)

28. J. Li, Y. Liu, W. Lei, Optimal operation for community-based multi-party microgrid in grid-connected and islanded modes. IEEE Trans. Smart Grid **9**(2), 756–765 (2018)

29. P.M. de Quevedo, J. Contreras, A. Mazza, G. Chicco, R. Porumb, Reliability assessment of microgrids with local and mobile generation, time-dependent profiles, and intraday reconfiguration. IEEE Trans. Ind. Appl. **54**(1), 61–72 (2017)

30. R. Ahmadi, A. Sheikholeslami, A. Nabavi Niaki, A. Ranjbar, Dynamic participation of doubly fed induction generators in multi-control area load frequency control. Int. Trans. Electr. Energy Syst. **25**(7), 1130–1147 (2015)

31. R. Ahmadiahangar, A. Sheykholeslami, Bulk virtual power plant, a novel concept for improving frequency control and stability in presence of large scale RES. Int. J. Mechatron. Electr. Comput. Technol. **4**(10), 1017–1044 (2014)

32. L. Guo, J. Su, J. Lai, Y. Wang, Research on power scheduling strategy for microgrid in islanding mode. Int. Trans. Electr. Energy Syst. **28**(2), e2493 (2018). https://doi.org/10.1002/etep.2493

33. A. Rahmoun, N. Beg, A. Rosin, H. Biechl, Stability and eigenvalue sensitivity analysis of a BESS model in a microgrid, in *2018 IEEE International Energy Conference (ENERGYCON)* (2018)

34. S. Choudhury, P. Bhowmik, P.K. Rout, Robust dynamic fuzzy-based enhanced VPD/FQB controller for load sharing in microgrid with distributed generators. Electr. Eng. **100**(4), 2457–2472 (2018)

35. S. Choudhury, P. Bhowmik, P.K. Rout, Seeker optimization approach to dynamic PI based virtual impedance drooping for economic load sharing between PV and SOFC in an islanded microgrid. Sustain. Cities Soc. **37**, 550–562 (2018)

36. M. Cucuzzella, G.P. Incremona, A. Ferrara, Decentralized sliding mode control of islanded ac microgrids with arbitrary topology. IEEE Trans. Industr. Electron. **64**(8), 6706–6713 (2017)

37. A. Bidram, A. Davoudi, F.L. Lewis, J.M. Guerrero, Distributed cooperative secondary control of microgrids using feedback linearization. IEEE Trans. Power Syst. **28**(3), 3462–3470 (2013)

38. W. Liu, W. Giu, W. Sheng, X. Meng, S. Xue, M. Chen, Pinning based distributed cooperative control for autonomous microgrids under uncertain communication topologies. IEEE Trans. Power Syst. **31**(2), 1320–1329 (2016)

39. S. Manaffam, M. Talebi, A.K. Jain, A. Behal, Intelligent pinning based cooperative secondary control of distributed generators for microgrid in islanding operation mode. IEEE Trans. Power Syst. **33**(2), 1364–1373 (2018)

40. Economic load sharing in a D-STATCOM integrated islanded microgrid based on fuzzy logic and seeker optimization approach. Sustain. Cities Soc. **37**, 57–69 (2018)

41. G.G. Talapur, H.M. Suryawanshi, A reliable microgrid with seamless transition between grid connected and islanded mode for residential community with enhanced power quality. IEEE Trans. Ind. Appl. **54**(5), 5246–5255 (2018)
42. L.G. Meegahapola, D. Robinson, A.P. Agalgaonkar, S. Perera, P. Ciufo, Microgrids of commercial buildings: strategies to manage mode transfer from grid connected to islanded mode. IEEE Trans. Sustain. Energy. **5**(4), 1337–1347 (2014)
43. Hybrid islanding detection with optimum feature selection and minimum NDZ. Int. Trans. Electr. Energy Syst. **28**(10), e2602 (2018)
44. A. Micallef, M. Apap, C. Spiteri-Staines, J.M. Guerrero, Single-phase microgrid with seamless transition capabilities between modes of operation. IEEE Trans. Smart Grid **6**(6), 2736–2745 (2015)
45. Y.A. Mohamed, A.A. Radwan, Hierarchical control system for robust microgrid operation and seamless mode transfer in active distribution systems. IEEE Trans. Smart Grid **2**(2), 352–362 (2016)
46. P. Bhowmik, S. Chandak, P. Kumar, State of charge and state of power management among the energy storage systems by the fuzzy tuned dynamic exponent and the dynamic PI controller. J. Energy Storage **19**, 348–363 (2018)
47. P. Bhowmik, S. Chandak, P.K. Rout, State of charge and state of power management in a hybrid energy storage system by the self-tuned dynamic exponent and the fuzzy-based dynamic PI controller. Int. Trans. Electr. Energy Syst. **29**(5), e2848 (2019). https://doi.org/10.1002/2050-7038.2848
48. M. Moretti, S.N. Djomo, H. Azadi, K. May, K. De Vos, S. Van Passel, N. Witters, A systematic review of environmental and economic impacts of smart grids. Renew. Sustain. Energy Rev **68**, 888–889 (2017)
49. M. Mahmudizad, R. Ahmadiahangar, Improving load frequency control of multi-area power system by considering uncertainty by using optimized type 2 fuzzy pid controller with the harmony search algorithm. World Acad. Sci. Eng. Technol. Int. J. Electr. Comput. Energ. Electron. Commun Eng. **10**(8), 1051–1061 (2016)
50. K.M. Tan, V.K. Ramachandaramurthy, J.Y. Yong, Integration of electric vehicles in smart grid: a review on vehicle to grid technologies and optimization techniques. Renew. Sustain. Energy Rev. **53**, 720–732 (2016)
51. S.V. Oprea, A. Bâra, G. Ifrim, Flattening the electricity consumption peak and reducing the electricity payment for residential consumers in the context of smart grid by means of shifting optimization algorithm. Comput. Industr. Eng. **122**, 125–139 (2018)
52. R. Ahmadiahangar, T. Häring, A. Rosin, T. Korõtko, J. Martins, Residential load forecasting for flexibility prediction using machine learning-based regression model, in *2019 IEEE International Conference on Environment and Electrical Engineering and 2019 IEEE Industrial and Commercial Power Systems Europe (EEEIC/I&CPS Europe)*, Genoa, Italy, pp. 2–19
53. N. Shabbir, R. Ahmadiahangar, L. Kütt, A. Rosin, Comparison of machine learning based methods for residential load forecasting, in *2019 Electric Power Quality and Supply Reliability Conference (PQ) & 2019 Symposium on Electrical Engineering and Mechatronics (SEEM)* (2019)
54. T. Häring, R. Ahmadiahangar, A. Rosin, H. Biechl, Impact of load matching algorithms on the battery capacity with different household occupancies, in *IECON 2019-45th Annual Conference of the IEEE Industrial Electronics Society*, Lisbon (2019)
55. VassaETT, Smart grid 2013 global impact report. SMARTGRID.GOV, DOE, U.S., (October 2013)
56. R. Kappagantu, S.A. Daniel, Journal of Electrical Systems and Information Technology **5**, 453–467 (2018)
57. C. Tu, X. He, Z. Shuai, F. Jiang, Big data issues in smart grid–A review. Renew. Sustain. Energy Rev. **79**, 1099–1107 (2017)
58. H. Daki, A. El Hannani, A. Aqqal, A. Haidine, A. Dahbi, Big Data management in smart grid: concepts, requirements and implementation. Big Data **4**(13), 1–19 (2017)

59. A.B. Dayani, H. Fazlollahtabar, R. Ahmadiahangar, A. Rosin, M.S. Naderi, M. Bagheri, Applying reinforcement learning method for real-time energy management, in *2019 IEEE International Conference on Environment and Electrical Engineering and 2019 IEEE Industrial and Commercial Power Systems Europe (EEEIC/I&CPS Europe)* (2019)

60. H. Fallahzadeh-Abarghouei, S. Hasanvand, A. Nikoobakht, Decentralized and hierarchical voltage management of renewable energy resources in distribution smart grid. Int. J. Electr. Power Energy Syst. **100**, 117–128 (2018)

61. K.G. Firouzjah, R. Ahmadiahangar, A. Rosin, T. Häring, A fast current harmonic detection and mitigation strategy for shunt active filter, in *2019 Electric Power Quality and Supply Reliability Conference (PQ) & 2019 Symposium on Electrical Engineering and Mechatronics (SEEM)* (2019)

62. A. Fernández-Guillamón, E. Gómez-Lázaro, E. Muljadi, Power systems with high renewable energy sources: a review of inertia and frequency control strategies over time. Renew. Sustain. Energy Rev. **115**, 109369 (2019)

Chapter 2
On the Concept of Flexibility in Electrical Power Systems: Signs of Inflexibility

2.1 Overview

Maintaining the balance between load and supply is one of the main challenges of the power system operation. This balance is perturbed by three types of events in different time frames: rapid random fluctuations, slow periodical fluctuations, and rare instantaneous changes [1]. Flexibility is the capability of the system to withstand variations originated from both generation and demand-sides. This chapter discusses the general concept of flexibility in the power system as well as main sources of flexibility and signs of inflexibility.

2.2 What is the Power System Flexibility?

Energy consumption is the most important factor of environmental concerns, and it plays an essential role in the economy. Increase in the penetration of renewable energy sources (RES), results in challenges in the Power Systems (PS). In this regard, the main challenge is the management of the increased variability and uncertainty imposed by large penetration of RES in the power balance [2].

Flexibility is a recent concept in PS and has been officially recognized by organizations like IEA [3] and NERC [4]. However, there is no universal definition for flexibility. Nevertheless, flexibility can be simply defined as the PS's ability to

© The Author(s), under exclusive license to Springer Nature Singapore Pte Ltd. 2020
R. Ahmadiahangar et al. *Demand-side Flexibility in Smart Grid*,
SpringerBriefs in Applied Sciences and Technology,
https://doi.org/10.1007/978-981-15-4627-3_2

respond to both expected and unexpected changes in demand and supply [5]. Recent studies in this area are focusing on planning, scheduling, and exploitation of flexibility mostly from generation units and large-scale energy storage systems [6, 7]. Another challenge is that while there are many emerging flexibility metrics and assessment methods, there is no standard metric for measuring flexibility to date, and metrics continue to evolve [5]. It is worth mentioning that it makes more economic sense to plan for the flexibility rather than exploring suboptimal investments after the flexibility issues arise in a PS [8].

The study in [9] estimates the flexibility of 12–23GW for the Northern European countries (Sweden, Denmark, Norway, Finland, Estonia, Latvia, and Lithuania). This would be 15–30% of the peak load of the region.

Figure 2.1 shows the potential of different options of flexibility in the following three operational time frames [10], where:

- **Short term flexibility** refers to balancing markets up to one hour,
  Traditionally thermal power plants are in this category.
- **Mid-term** flexibility refers to day-ahead and daily markets.
- **Long term** flexibility refers to planning phase.

In Fig. 2.1, color tense shows the suitability of the flexibility option with the timeframes. Also, bold/underscore titles present large scale and mature technologies while red colors present small-scale ones. Flexibility resources are divided to three categories of Supply, Demand and Storage in this research.

The study in [11] classifies flexibility resources chronologically as shown in Fig. 2.2. Operational time frames are defined as Long-term, Mid-term, Short-term and Super short term in this study.

In 2018, IEA provided a revised definition of flexibility as: *"all relevant characteristics of a power system that facilitates the reliable and cost-effective management of variability and uncertainty in both supply and demand"* [12].

The International Renewable Energy Agency (IRENA) defined flexibility in 2018 as

> the capability of a power system to cope with the variability and uncertainty that VRE (variable renewable energy) generation introduces into the system in different timescales, from the very short to the long term, avoiding curtailment of VRE and reliably supplying all the demanded energy to customers. [13]

Flexibility can be provided to the grid by different parts of the power system, as well as different technologies. The study in [14] presents different sources of flexibility as

- **Demand-side flexibility**, demand-side is generally known as a wide source for providing flexibility. The interesting fact about demand-side flexibility is that it usually does not need a high initial investment cost. Smart grid is also enabling new opportunities to exploit flexibility from the demand-side [15–17].

Generally, demand-side flexibility is obtained through different demand response programs [18, 19]. These programs can be either incentive-based or price-based.

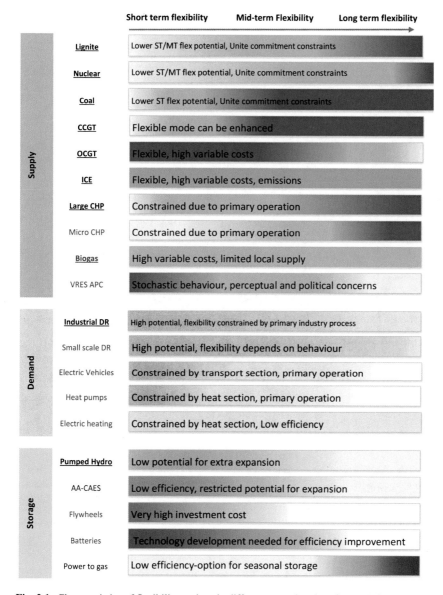

Fig. 2.1 Characteristics of flexibility options in different operation time frames [10]

Incentive-based programs change the amount of consumption and include direct load control, curtailable load services, demand bidding or buyback programs, and emergency DR among others, while price-based programs change consumption pattern and mainly include time-of-use (ToU), critical peak pricing (CPP), peak time rebate (PTR), and real-time pricing (RTP) programs. Figure 2.3 illustrates the possible exploitations of flexibility from the demand-side.

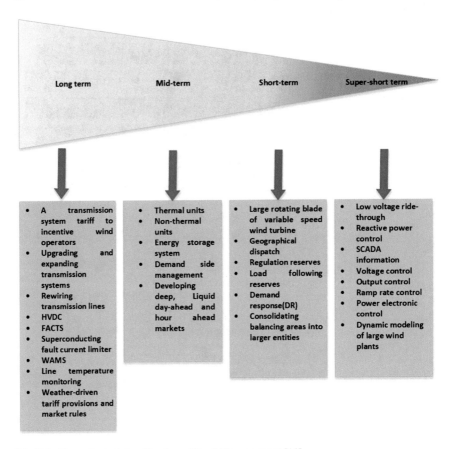

Fig. 2.2 Chronological classification of flexibility resources [11]

IEA estimates that nearly 185 GW of demand-side flexibility could be reached cost-effectively by 2040 [9]. Globally, IEA estimates 15% flexibility from the total demand.

- **Supply-side flexibility**, traditionally, conventional generation units provide flexibility and ramping capability for the grid. Depending on their levels of flexibility, power plants are classified into baseload (like coal and nuclear), peaking and load-following regimes (gas and hydropower) [20]. Load-following generations are the ones providing balancing power and flexibility at the moment.

The power outputs from RES are subject to high-level uncertainty, which naturally decrease grid flexibility. Meanwhile, researches are going on developing strategic curtailment of RES to increase flexibility in the power systems [20].

- **Grid side flexibility**, Grid reconfiguration [21], smartification, dynamic line ratings, dynamic grid reconfiguration, wide-area interconnections [22], and meshed operations can be options for providing flexibility from the grid side. Main

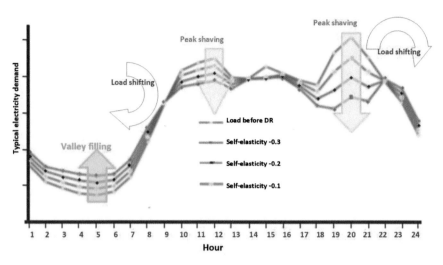

Fig. 2.3 An illustrative example of Flexibility from Demand Response Programs [14]

aim is to balance supply and demand over larger balancing areas (Aggregator/DSO/TSO/BRP), as well as cross-border interconnections to enable the trade of flexibility between different countries. It also refers to the existence of advanced controls to enhance communication among system elements that enables, for example, automated control of generators, automatic activation of demand response or advanced power flow control (MTDC, HVDC, FACTS).

- **Energy storage systems (ESS)** are probably the most common solution to compensate variable generation of RES and provide flexibility in the smart grid. ESS technology can be divided into five main categories of physical (compressed air and pumped hydro), electromechanical (flywheels), electrochemical (fuel cells and batteries) [23], electrostatic (capacitors), and electromagnetic storages (superconducting magnets) [14]. ESS technologies are becoming more viable in providing flexibility in bulk quantities because of their technology development and falling capital costs. Flexibility characteristics of energy storage systems for different time frames are shown in Fig. 2.4 [10].

- **Integrated Energy System (IES)**, formed by integrating and optimizing multiple energy systems, provides a new solution to energy and environmental problems and believed to add more dimensions to flexibility. The system-wide benefit of an IES can be realized through optimally coordinated multistage planning. A local energy system model, such as for a microgrid, requires a higher level of details in components and operation strategy. The coordinated optimal planning model must cover several energy systems with multiple energy carriers such as the power system, natural gas system, intelligent transportation system, and district heating system, and the interactions among them. For example, the consequence of local investment in renewable generation can be reflected in the district heating system and vice versa.

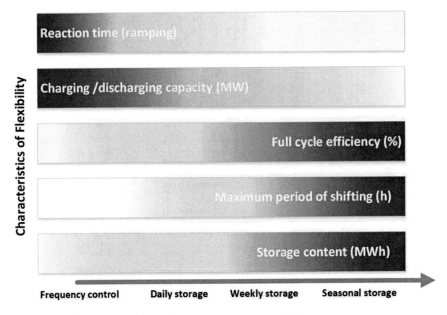

Fig. 2.4 Flexibility characteristics of energy storage systems [10]

- **Energy markets** besides physical or technological means, properly designed markets can also increase the flexibility of the power system. Especially in the demand-side, market is the main enabler for flexibility exploitation. Recently, some research are going on proposing flexibility markets [24, 25]. Electricity grid tariffs are also introduced in [26] as a tool for increasing flexibility in Danish system.

2.3   Signs of Inflexibility

Based on the study in [5], signs of inflexibility include the following.

2.3.1   Challenges in Demand-Supply Balancing

Probably the first sign of inflexibility is the inability of the grid to maintain frequency. Conventionally power systems are synchronized to the grid and their stored kinetic energy is automatically extracted in response to a sudden power imbalance system demand-supply balancing depends on synchronous machines connected to the grid. However, the power systems generation fleet is changing RES [27]. Therefore, the

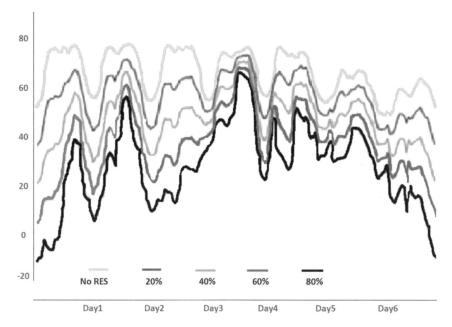

Fig. 2.5 Daily pattern of net electricity demand with different penetration level of RES in Germany [10]

situation gets even worth. Finally, this will lead to frequency excursions or dropped load [28].

Figure 2.5 shows the daily pattern of net electricity demand with different penetration levels of RES in Germany. As it is shown, net demand variations increase noticeably with the increase in penetration level of RES.

2.3.2 Significant Renewable Energy Curtailments

This is the case when there is an overgeneration in the grid for long periods, mostly in nights or certain seasons. The reasons are usually the following:

- **System-wide oversupply**

    Usually happens when there is simply not enough demand for all the renewable electricity that is available. This is not a concern at low levels of curtailments, but would be problematic in higher levels.

- **Transmission grid constraints**

    Transmission grid constraints happen when the share of renewable sources electricity generation is high in a local area and there is insufficient transmission infrastructure to deliver the electricity to a place where it could be used; this is so called energy curtailment, which is a common problem of transmission grids all over the world [29–31]; this is so called energy curtailment, which is a common problem of transmission grids all over the world. An example of energy curtailment, in 2018, the total wind energy generated in Ireland and Northern Ireland was 11,076 GWh, while 707 GWh of wind energy was dispatched-down [32]. This represents 6% of the total available wind energy in 2018 and is an increase of 321 GWh on the 2017 value. In Ireland, the dispatch down energy from wind resources was 457 GWh. This is equivalent to 5% of the total available wind energy.

2.3.3 Area Balance Violations

This is referred to the deviations from the scheduled area power balance and can indicate how frequently a system cannot meet its electricity balancing responsibility [5]. Area balance violation is an extreme sign of insufficient flexibility in the grid.

2.3.4 Negative Market Prices

Recently, renewables generations have made negative prices increasingly common in the electricity market. In this situation there is no possible solution for purchasing excess energy and it leads to negative electricity price. This is a clear sign of the lack of sufficient available flexibility in the grid and means that the power suppliers have to pay their wholesale customers to buy electric energy.

2.3.5 Price Volatility

Bulk integration of RES in the power system causes variations in electricity price with different patterns, generally called price volatility. Although this is still a contrversial issue in the economic literature, some studies [33] present evident for price fluctuation and volatility dynamics.

Acknowledgements This work has been supported by the European Commission through the H2020 project Finest Twins (grant No. 856602).

References

1. D. Kirschen, G. Strback, *Fundamentals of Power System Economics* (Wiley Ltd, Chichester, England, 2004)
2. H. Nosair, F. Bouffard, Flexibility envelopes for power system operational planning. IEEE Trans. Sustain. Energy **6**(3), 800–809 (2015)
3. International Energy Agency, in *Harnessing Variable Renewables: A Guide to the Balancing Challenge*, OECD/IEA, Paris, France, Tech. Rep. 61-2011-17-1P1 (2011)
4. North American Electric Reliability Corporation, *Accomodating High Levels of Variable Generation*, NERC, Princeton, NJ, USA, Tech. Rep (2009)
5. J. Cochran, M. Miller, O. Zinaman, M. Milligan, D. Arent, B. Palmintier, *Flexibility in 21st Century Power Systems*. NREL (2018)
6. H. Ji, C. Wang, P. Li, G. Song, Quantified analysis method for operational flexibility of active distribution networks with high penetration of distributed generators. Appl. Energy **239**, 706–714 (2019)
7. H. Jiahua, M.R. Sarker, J. Wang, F. Wen, W. Liu, Provision of flexible ramping product by battery energy storage in day-ahead energy and reserve markets. IET Gener. Transm. Distrib. **12**(10), 2256–2265 (2018)
8. E. Taibi, T. Nikolakakis, L. Gutierrez, C. Fernandez, *Power System Flexibility for the Energy Transition, Overview for Policy Makers* (International Renewable Energy Agency, Abu Dhabi, 2018)
9. L. Söder, P.D. Lund, H. Koduvere, T.F. Bolkesjø, G.H. Rossebø, E. Rosenlund-Soysal, K. Skytte, J. Katz, D. Blumberga, A review of demand side flexibility potential in Northern Europe. Renew. Sustain. sEnergy Rev. **91**, 654–664 (2018)
10. G. Papaefthymiou, K. Grave, K. Dragoon, *Flexibility Options in Electricity Systems*, Ecofys (2014)
11. M.I. Alizadeh, M.P. Moghaddam, N. Amjady, P. Siano, M.K. Sheikh-El-Eslami, Flexibility in future power systems with high renewable penetration: a review. Renew. Sustain. Energy Rev. **57**, 1186–1193 (2016)
12. IEA, *Status of Power System Transformation 2018: Advanced Power Plant Flexibility* (IEA, Paris, 2018)
13. International Renewable Energy Agency (IRENA), *Power System Flexibility for the Energy Transition, Part 1: Overview for Policy makers* (IRENA, Abu Dhabi, 2018)
14. M.R. Cruz, D.Z. Fitiwi, S.F. Santos, J.P. Catalão, A comprehensive survey of flexibility options for supporting the low-carbon energy future. Renew. Sustain. Energy Rev. **97**, 338–353 (2018)
15. N. Shabbir, R. Ahmadiahangar, Lauri Kütt, Muhamamd N Iqbal, Argo Rosin, "Forecasting Short Term Wind Energy Generation using Machine Learning," in *2019 IEEE 60th International Scientific Conference on Power and Electrical Engineering of Riga Technical University (RTUCON)*, 2019
16. T. Häring, R. Ahmadiahangar, A. Rosin, H. Biechl, T. Korõtko, Comparison of the impact of different household occupancies on load matching algorithms in *2019 Electric Power Quality and Supply Reliability Conference (PQ) & 2019 Symposium on Electrical Engineering and Mechatronics (SEEM)* (2019)
17. A.B. Dayani, H. Fazlollahtabar, R. Ahmadiahangar, A. Rosin, M.S. Naderi, M. Bagheri, Applying reinforcement learning method for real-time energy management, in *2019 IEEE International Conference on Environment and Electrical Engineering and 2019 IEEE Industrial and Commercial Power Systems Europe (EEEIC/I&CPS Europe)* (2019)
18. R. AhmadiAhangar, A. Rosin, A.N. Niaki, I. Palu, T. Korõtko, A review on real-time simulation and analysis methods of microgrids. Int. Trans. Electr. Energy Syst **29**(11), e12106 (2019)
19. R. Ahmadiahangar, T. Häring, A. Rosin, T. Korõtko, J. Martins, Residential load forecasting for flexibility prediction using machine learning-based regression model, in *2019 IEEE International Conference on Environment and Electrical Engineering and 2019 IEEE Industrial and Commercial Power Systems Europe (EEEIC/I&CPS Europe)*, Genoa, Italy (2019)

20. P.D. Lund, J. Lindgren, J. Mikkola, E.J. Salpakari, Review of energy system flexibility measures to enable high levels of variable renewable electricity. Renew. Sustain. Energy Rev. **45**, 785–807 (2015)

21. K. Peterson, R. Ahmadiahangar, N. Shabbir, T. Vinnal, Analysis of microgrid configuration effects on energy efficiency, in *2019 IEEE 60th International Scientific Conference on Power and Electrical Engineering of Riga Technical University (RTUCON)* (2019)

22. M. Mahmudizad, R.A. Ahangar, Improving load frequency control of multi-area power system by considering uncertainty by using optimized type 2 fuzzy pid controller with the harmony search algorithm. World Acad. Sci. Eng. Technol. Int. J. Electr. Comput. Energ. Electron. Commun. Eng. **10**(8), 1051–1061 (2016)

23. T. Häring, R. Ahmadiahangar, A. Rosin, H. Biechl, Impact of load matching algorithms on the battery capacity with different household occupancies, in *IECON 2019-45th Annual Conference of the IEEE Industrial Electronics Society* (Lisbon, 2019)

24. S. Goutte, P. Vassilopoulos, The value of flexibility in power markets. Energy Policy **125**, 347–357 (2019)

25. D. Schwabeneder, A. Fleischhacker, G. Lettner, H. Auer, Assessing the impact of load-shifting restrictions on profitability of load flexibilities. Appl. Energy **1**(255), 113860 (2019)

26. C. Bergaentzlé, I.G. Jensen, K. Skytte, O.J. Olsen, Electricity grid tariffs as a tool for flexible energy systems: A Danish case study. Energy Policy **126**, 12–21 (2019)

27. A. Fernández-Guillamón, E. Gómez-Lázaro, E. Muljadic, Á. Molina-Garcí, Power systems with high renewable energy sources: a review of inertia and frequency control strategies over time. Renew. Sustain. Energy Rev. **109369**, 115 (2019)

28. N. Shabbir, R. Ahmadiahangar, L. Kütt, A. Rosin, Comparison of machine learning based methods for residential load forecasting, in *2019 Electric Power Quality and Supply Reliability Conference (PQ) & 2019 Symposium on Electrical Engineering and Mechatronics (SEEM)* (2019)

29. M. Specht, https://blog.ucsusa.org/mark-specht/renewable-energy-curtailment-101, Union of Concerned Scientist, 25 June 2019. [Online]. Accessed 10 July 2019

30. R. Ahmadi, A. Sheikholeslami, A. Nabavi Niaki, A. Ranjbar, Dynamic participation of doubly fed induction generators in multi-control area load frequency control. Int. Trans. Electr. Energy Syst. **25**(7), 1130–1147 (2015)

31. R. Ahmadi, F. Ghardashi, D. Kabiri, A. Sheykholeslami, H. Haeri, Voltage and frequency control in smart distribution systems in presence of DER using flywheel energy storage system. *IET Digital Library* (2013)

32. *Annual Renewable Energy Constraint and Curtailment Report 2018* (EIRGrid, May 2019)

33. G. Wang, S. Zheng, J. Wang, Fluctuation and volatility dynamics of stochastic interacting energy futures price model. Physica A **537**, 1 (2020)

Chapter 3
Investigating Different Sources of Flexibility in Power System

3.1 Overview

Increasing the integration of variable renewable energy sources to power systems has a negative effect on the reliable operation of the grid. To overcome this challenge, increasing flexibility is known to be the main key. This chapter aims to provide a comprehensive analysis of different sources of flexibility in power systems.

3.2 Challenges in Supply-Side Flexibility

Fossil fuels (including coal, oil, and gas) are still the most important energy source for electricity production in the EU. In addition, some of the European Economic Area (EEA) countries still depend on nuclear power as a prevalent energy source. Despite the latest high growth rates in the wind and solar power in a few member states, Variable Renewable Energy Resources (VRERs) contribution in energy generation in some EEA countries is not considerable or there is more place for the growth of VRERs. In some areas, relatively sizeable hydropower plants have remained as major Renewable Energy Resources (RERs). Implementation of Combined Heat and Power (CHP), in spite of important advances in some member states, has remained low in comparison to the EU target for VRERs penetration.

Traditionally, power system flexibility is provided by the supply-side. Since variable renewable energy, resources are displacing traditional supply-side flexibility providers, and at the same time, increasing power generation variations, it can be concluded that supply-side flexibility is decreasing noticeably. While searching for new options of flexibility, increasing available flexibility of supply-side must be in high priority. Figure 3.1 shows the need for flexibility with increasing share of RES.

© The Author(s), under exclusive license to Springer Nature Singapore Pte Ltd. 2020
R. Ahmadiahangar et al. *Demand-side Flexibility in Smart Grid*,
SpringerBriefs in Applied Sciences and Technology,
https://doi.org/10.1007/978-981-15-4627-3_3

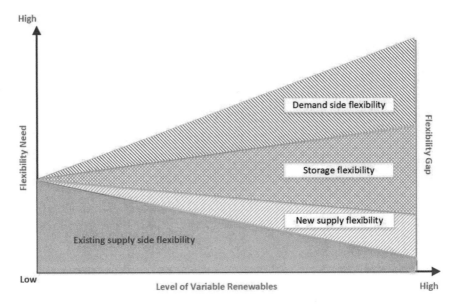

Fig. 3.1 The emerging flexibility gap [1]

Supply-side flexibility can be divided into two main categories of conventional power plants and renewable energy sources as follows.

3.2.1 Conventional Power Plants

Conventional power plants including coal, gas, oil, oil shell, and CHP are the most well-known source of flexibility in power system.

Most significant parameters to consider in the modeling of conventional power plants flexibility are as follows:

- Ramp rate
- Maximum generation
- Minimum generation
- Part-load operation.

Gas turbines and internal combustion units are the most flexible ones [2].

3.2.2 Renewable Energy Sources

Main differences of power generation from renewable energy sources and conventional power plants can be counted as follows:

- RES production (mostly solar and wind generation) depends on weather conditions; therefore, it is highly unpredictable, intermittent and dependent on the accuracy of forecast methods [3]. At high share of RES, generation-side might have sever fluctuations [4, 5].
- RES operation cost is considerably low comparing to conventional power plants [6].
- RES output, show different levels of short-term, medium-term (daily, weekly) and long-term (seasonal) variations [7, 8].

Another effect of RES on power system is decreasing total inertia and as a result changing the response time. In this regard, fast RES response and spare generation capacity are needed to tackle RES volatility [9, 10].

Overall, most significant parameters to consider in modeling of RES flexibility are:

- The power of renewable energy can be considered as negative flexibility in generation part for power system.

3.3 Quantifying Grid Side Flexibility

In the power system, from the grid side, innovative technologies, communication and monitoring possibilities, and increased interaction and information exchange are enablers to provide flexibility to the grid. Meanwhile, methods and procedures related to system planning and operation will be required to utilize available flexibility to provide the most value to society.

Regarding [11] flexibility in grid side can be categorized into three parts.

3.3.1 Flexibility for Power

Short-term equilibrium between power supply and power demand, a system-wide requirement for maintaining the frequency stability.

Main Rationale: Increased amount of intermittent, weather dependent, power supply in the generation mix.

Activation Timescale: Fractions of a second up to an hour [11].

3.3.2 Flexibility for Energy

Medium to long-term equilibrium between energy supply and energy demand, a system-wide requirement for demand scenarios over time.

Main Rationale: Decreased amount of fuel storage-based energy supply in the generation mix.

Activation Timescale: Hours to several years [11].

3.3.3 Flexibility for Transfer Capacity

Need Description: Short to medium-term ability to transfer power between supply and demand, where local or regional limitations may cause bottlenecks resulting in congestion costs.

Main Rationale: Increased utilization levels, with increased peak demands and increased peak supply.

Activation Timescale: Minutes to several hours [11].

3.4 Flexibility for Voltage

Need Description: Short-term ability to keep the bus voltages within predefined limits, local and regional requirements.

Main Rationale: Increased amount of distributed power generation in the distribution systems, resulting in bi-directional power flows and increased variance of operating scenarios.

Activation Timescale: Seconds to tens of minutes [11].

Examples of flexibility solutions for each category with implementation levels from local to system-wide are shown in Fig. 3.2.

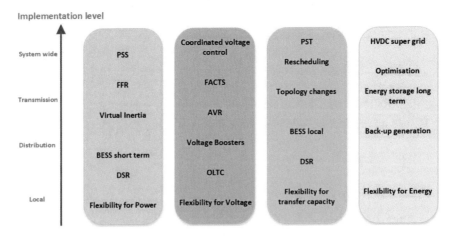

Fig. 3.2 Examples of flexibility solutions for each category with implementation levels [11]

3.5 Exploiting Demand-Side Flexibility (DSF)

Load management, also known as demand-side management (DSM) or load matching, is the process of balancing the supply of electricity by adjusting or controlling the electrical load rather than the power generation. In nZEB, the consumption is not homogenous. Different appliances have different regimes, priorities, and roles [12, 13].

Usage of appliances, which depend less on customer behavior and habits, are more flexible regarding energy management. The flexibility refers to physical characteristics of loads like energy storage or scheduling capability. Load management in terms of time is closely related to customer's needs or the convenience and depends on the functional possibilities, technical characteristics, and surrounding environment (including building construction) [14]. By flexibility, most loads in nZEB can be divided into three main priority groups [15–17]:

I.   flexible loads, e.g., water heaters, dishwashers, washing machines, and HVAC;
II. almost flexible loads, e.g., refrigerators, boiling kettles, coffee machines, floor heating, irons, and vacuum cleaners;
III. non-flexible loads, e.g., TV sets, PCs with a modem, home cinema, and music centers, cooking stoves, kitchen ventilation, bathroom lighting, and ventilation.

Also in commercial buildings (i.e., office buildings), 50% of total consumption has scheduling capability and flexibility to control at least an hour. Flexible loads in commercial buildings are heating, ventilation, and air conditioning (HVAC). Figures 3.3 and 3.4 show the distribution of electricity consumption on a typical apartment or office building.

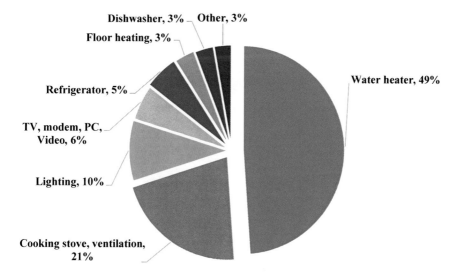

Fig. 3.3 Distribution of electrical consumption in typical apartment/household

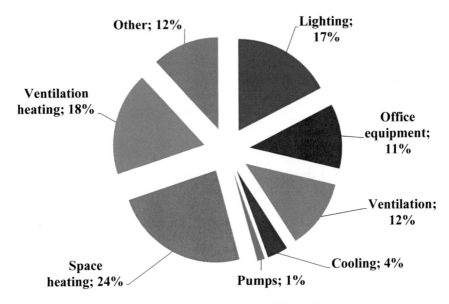

Fig. 3.4 Distribution of electrical consumption in typical office building

Load management algorithms can be divided into two main modes, on-grid (grid-connected) and off-grid (island mode) mode, which loads management on the building level, has different purposes in each one. Figure 3.5 shows load management algorithms classification.

In on-grid mode, the goal of most control algorithms is achieving cost reduction in utility or customer side, while power quality and supply reliability are usually guaranteed by the utility. These algorithms are often called market pricing or price-based control algorithms [18]. While most of the market pricing algorithms have the indirect goal to support the grid, we classify these control algorithms as market pricing

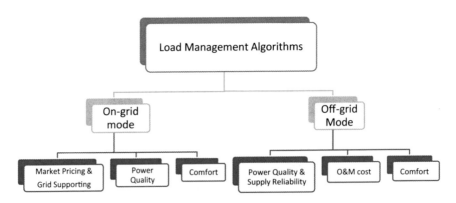

Fig. 3.5 Classification of load management algorithms

and grid supporting algorithms. In weak on-grid systems, also the power quality based control systems [19, 20] and algorithm have an important role to guarantee system work according to EN50160 and other power quality standards. Third import algorithms in on-grid mode are comfort-based control algorithms, which serve the interests of the customer to provide them a highest possible comfort level. These types of control algorithms are mainly used in home automation systems [21]. Often are used a combination of described algorithms. Classification of these algorithms depends on the priorities of grid and customers [22, 23].

In off-grid mode, the goal of most control algorithms is achieving power quality and supply reliability on microgrid (building or district) level. Another and not less important goal of control algorithms are the reduction of operational costs in off-grid mode [24]. To reduce the running costs of zero energy building, it is important to reduce the operational and maintenance (O&M) costs of power generation and storage systems in off-grid mode. Reduction of O&M costs is achieved by optimized power flow control in building taking into account the parameters and processes, which could affect the system aging and profitability. Similar to on-grid mode, the third type of control algorithms are in off-grid mode the customer's comfort. In off-grid mode, customer comfort is highly dependent on renewable and storage systems scaling and optimization.

On-grid mode control algorithms are mainly based on cost reduction, while power quality and supply reliability are usually guaranteed by utility. In this case, customer's engagement in DR program can be achieved through motivating them by priced based or intensive based algorithms and programs developed by utility in order to maintain the grid and reduce gird side costs [25] Classification of market pricing and grid supporting algorithms are shown in Fig. 3.6.

In grid-connected nZEB the flexible loads are defined as loads that can be shifted from a high tariff period to a low tariff period without any investments to additional

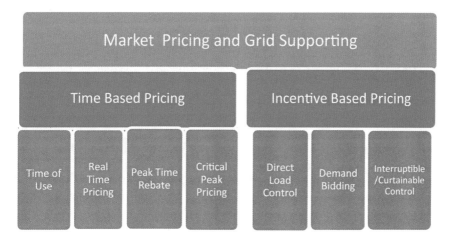

Fig. 3.6 Market pricing and grid supporting algorithms and programs in On-grid mode [18]

electrical or thermal energy storage. Depending on the electricity market prices the set point for the thermostatic control will be calculated for each time step. In on-grid systems with the combination of the optimum price-based control algorithms, it is possible to save about 20% of the costs for the appliances' energy consumption. Depending on amount of flexible load in the building, it could be also 10–20% of the electrical energy costs of the apartment. [26] reported that HEMS could reduce the operational cost of electricity by 23.1% on average, or reduce residential peak demand by 29.6%.

The coordination between different priced based DR programs of nZEBs is vital to maintain the grid efficiency. Otherwise, in the case of non-coordinated programs, new peak hours are likely to be formed when electricity prices are low [27, 28].

Beside flexible loads, Plug-in Hybrid Electrical Vehicles (PHEV) are penetrating into nZEBs in recent years. Since these vehicles have batteries that can be charged at different levels by the grid and can be discharged to return the energy back to the grid (e.g., vehicle to grid capability), it is necessary to incorporate PHEV in the load management procedure [29].

In off-grid mode, the most important challenge is to maintain power quality and supply reliability by balancing of renewable energy sources (PV-systems or wind turbines), with additional controllable power generation and efficient control of energy storage(s) (typically a battery) and flexible loads in nZEB [30–32] European countries have made it compulsory to use certain amounts for the minimum share of renewable resources in nZEBs [33]. High power fluctuation of local renewable systems is the main cause of reduced voltage quality in off-grid mode nZEB which leads to decreasing lifetime of electrical devices.

Economically most feasible solution for balancing renewables is control of flexible load. Load management according to renewable systems power generation is often called as load matching. Making use of thermal storages that are already in the household can, therefore, be a useful addition to generate a more efficient off-grid system. By using flexible loads for load matching is possible to reduce the install capacitance of the battery, hence reducing the investment cost of balancing (control) system. Decreasing battery capacitance results in cost reduction of entire system. Another option for better load matching is to utilize available thermal storages and heat storage capability of nZEB. As a result, by using load matching the battery storage could be reduced up to 30%. This enables a large reduction of initial investment costs in an electrical storage system. Besides investment, operation and maintenance costs may also be reduced with optimal determination of charge/discharge state of battery energy storage inside nZEB, including PHEV.

An accurate load forecast and scheduling will have an impressive effect on cost reduction and peak shaving in both on/off-grid modes. Mathematical and heuristic optimization techniques are both used in the literature to improve schedulings drawback is they are commonly time-consuming when solving complex optimization problems. To overcome this, heuristic optimization techniques are increasingly attract attentions [34].

In most HEMS on/off-grid mode, user intervention is needed, to manually set customer-driven load priorities and comfort preferences. For making the HEMS more convenient and encouraging customers to participate in DR, it is necessary to autonomously and meta-heuristically schedule household appliances without user intervention, while considering occupants' desired comfort-level and lifestyles [35, 36].

Three main descriptors for DSF characteristics are as follows:

- shedability;
- controllability;
- acceptability.

For commercial buildings, these are defined as follows:

(1) Shedability is the theoretical potential for load shed or shifts for given end-use, which is associated with specific demand-side strategies assuming adequate communications and control capabilities.
(2) Controllability refers to a portion of load shed or shifts for given end-use, which is associated with equipment having in place required communications and controls capabilities for grid support activities.
(3) Acceptability is the portion of load and service compromise that end-users may be willing to accept as part of a consequence for providing DSF.

While the common characterization method of the energy flexibility is considering it as a static function in all timescale, the validity of this approach is questionable because energy-based systems are never at steady state. In addition, few researches characterize energy flexibility as a dynamic function [37].

Acknowledgements This work has been supported by the European Commission through the H2020 project Finest Twins (grant No. 856602).

References

1. *Ecofys, Flexibility Options in Electricity Systems, 2014* (2014). [Online]. Available: http://www.ecofys.com/files/files/ecofys-eci-2014-flexibility-options-in-elecelectricity-systems.pdf. Accessed 2019
2. G. Papaefthymiou, K. Grave, K. Dragoon, Flexibility options in electricity systems. Ecofys (2014)
3. N. Shabbir, R. Ahmadiahangar, L. Kütt, A. Rosin, Comparison of machine learning based methods for residential load forecasting, in *Electric Power Quality and Supply Reliability Conference (PQ) & 2019 Symposium on Electrical Engineering and Mechatronics (SEEM)* (2019)
4. M. Mahmudizad, R. Ahmadiahangar, Improving load frequency control of multi-area power system by considering uncertainty by using optimized type 2 fuzzy pid controller with the harmony search algorithm. World Acad. Sci. Eng. Technol. Int. J Electr. Comput. Energ. Electron. Commun. Eng. **10**(8), 1051–1061 (2016)
5. R. Ahmadiahangar, F. Ghardashi, D. Kabiri, A. Sheykholeslami, H. Haeri, Voltage and frequency control in smart distribution systems in presence of DER using flywheel energy storage system, in *22nd International Conference and Exhibition on Electricity Distribution, CIRED* (Stockholm, Sweden, 2013)

6. R. Ahmadi, A. Sheikholeslami, A.N. Niaki, A. Ranjbar, Dynamic participation of doubly fed induction generators in multi-control area load frequency control. Int. Trans. Electr. Energy Syst. **25**(7), 1130–1147 (2015)
7. R. AhmadiAhangar, A. Rosin, A.N. Niaki, I. Palu, T. Korõtko, A review on real-time simulation and analysis methods of microgrids. Int. Trans Electr. Energy Syst. **29**(11), e12106 (2019)
8. H. Karimi-Davijani, A. Sheikholeslami, R. Ahmadi, H. Livani, Active and reactive power control of DFIG using SVPWM converter, in 43rd International Universities Power Engineering Conference, UPEC, Padova, Italy (2008)
9. M.B. Anwar, M.S.E. Moursi, W. Xiao, Dispatching and frequency control strategies for marine current turbines based on doubly fed induction generator. IEEE Trans. Sustain. Energy **7**, 262–270 (2016)
10. Y. Wang, V. Silva, M. Lopez-Botet-Zulueta, Impact of high penetration of variable renewable generation on frequency dynamics in the continental Europe interconnected system. IET Renew. Power Gener. **10**(1), 10–16 (2016)
11. E. Hillberg, A. Zegers, B. Herndler, S. Wong, J. Pompee, *Power Transmission & Distribution Systems; Flexibility needs in the future power system*, ISGAN Annex 6 Power T&D Systems (2019)
12. P. Kadar, ZigBee controls the household appliances, in *International Conference on Intelligent Engineering Systems, INES,* pp. 189–192 (2009)
13. R. Ahmadiahangar, T. Häring, A. Rosin, T. Korõtko, J. Martins, Residential load forecasting for flexibility prediction using machine learning-based regression model, in *2019 IEEE International Conference on Environment and Electrical Engineering and 2019 IEEE Industrial and Commercial Power Systems Europe (EEEIC/I&CPS Europe)* (2019)
14. T. Häring, R. Ahmadiahangar, A. Rosin, H. Biechl, Impact of load matching algorithms on the battery capacity with different household occupancies, in *IECON 2019–45th Annual Conference of the IEEE Industrial Electronics Society* (Lisbon, 2019)
15. A. Rosin, H. Hõimoja, T. Möller, M. Lehtla, Residential electricity consumption and loads pattern analysis, in *Proceedings of the 2010 Electric Power Quality and Supply Reliability Conference* (Kuressaare, 2010)
16. I. Drovtar, J. Niitsoo, A. Rosin, J. Kilter, I. Palu, Electricity consumption analysis and power quality monitoring in commercial buildings, in *2012 Electric Power Quality and Supply Reliability* (Tartu, 2012)
17. A. Rosin, A. Auvaart, D. Lebedev, Analysis of operation times and electrical storage dimensioning for energy consumption shifting and balancing in residential areas. Elektronika Ir Elektrotechnika **120**(4), 15–20 (2012). https://doi.org/10.5755/j01.eee.120.4
18. X. Yan, Y. Ozturk, H. Zechun, Y. Song, A review on price-driven residential demand response. Renew. Sustain. Energy Rev. **96**, 411–419 (2018)
19. V.M. Lopez-Martin, F.J. Azcondo, A. Pigazo, Power quality enhancement in residential smart grids through power factor correction stages. IEEE Trans. Industr. Electron. **65**(11), 8553–8564 (2018)
20. A. Abessi, V. Vahidinasab, M.S. Ghazizadeh, Centralized support distributed voltage control by using end-users as reactive power support. IEEE Trans. Smart Grid **7**(1), 178–188 (2016)
21. X. Jina, K. Bakera, D. Christensen, S. Isley, Foresee: a user-centric home energy management system for energy efficiency and demand response. Appl. Energy **205**, 1583–1595 (2017)
22. A. Khan, N. Javaid, M.I. Khan, Time and device based priority induced comfort management in smart home within the consumer budget limitation. Sustain. Cities Soc. **41**, 538–555 (2018)
23. R. Kadavil, S. Lurb, S. Suryanarayanan, P.A. Aloise-Young, S. Isley, D. Christensen, An application of the analytic hierarchy process for prioritizing user preferences in the design of a home energy management system. Sustain. Energy Grids Netw. **16**, 196–206 (2018)
24. X. Wu, X. Hu, X. Yin, S.J. Moura, Stochastic optimal energy management of smart home with PEV energy storage. IEEE Trans. Smart Grid **9**(3), 2065–2071 (2018)
25. N.U. Hassan, Y.I. Khalid, C. Yuen, W. Tushar, Customer engagement plans for peak load reduction in residential smart grids. IEEE Trans. Smart Grid **6**(6), 3029–33041 (2015)

26. M. Beaudin, H. Zareipour, Home energy management systems: a review of modelling and complexity. Renew. Sustain. Energy Rev. **45**, 318–335 (2015)
27. C. Fan, G. Huang, Y. Sun, A collaborative control optimization of grid-connected net zero energy buildings for performance improvements at building group level. Energy **164**, 536–549 (2018)
28. Y. Tang, Q. Chen, J. Ning, Q. Wang, S. Feng, Y. Li, Hierarchical control strategy for residential demand response considering time-varying aggregated capacity. Electr. Power Energy Syst. **97**, 165–173 (2018)
29. S. Shafiee, M. Fotuhi-Firuzabad, M. Rastegar, Investigating the impacts of plug-in hybrid electric vehicles on power distribution systems. IEEE Trans. Smart Grid **4**(3), 1351–1360 (2013)
30. N. Shabbir, R. Ahmadiahangar, L. Kütt, M.N. Iqbal, A. Rosin, Forecasting short term wind energy generation using machine learning, in *2019 IEEE 60th International Scientific Conference on Power and Electrical Engineering of Riga Technical University (RTUCON)*, pp. 1–4 (IEEE, 2019)
31. T. Häring, R. Ahmadiahangar, A. Rosin, H. Biechl, T. Korõtko, Comparison of the impact of different household occupancies on load matching algorithms, in *2019 Electric Power Quality and Supply Reliability Conference (PQ) & 2019 Symposium on Electrical Engineering and Mechatronics (SEEM)*, pp. 1–6 (IEEE, 2019)
32. A.B. Dayani, H. Fazlollahtabar, R. Ahmadiahangar, A. Rosin, M.S. Naderi, M. Bagheri, Applying reinforcement learning method for real-time energy management, in *2019 IEEE International Conference on Environment and Electrical Engineering and 2019 IEEE Industrial and Commercial Power Systems Europe (EEEIC/I&CPS Europe)*, pp. 1–5 (IEEE, 2019)
33. S. Attiaa, P. Eleftherioub, F. Xenib, R. Morlotc, C. Ménézod, V.M. Kostopoulose, M. Betsie, Overview and future challenges of nearly zero energy buildings(nZEB) design in Southern Europe. Energy Build. **155**, 439–458 (2017)
34. H. Shareef, M.S. Ahmed, Review on home energy management system considering demand responses, smart technologies, and intelligent controllers. *Special Section on Energy Management in Buildings*. https://doi.org/10.1109/access.2018.2831917 (2018)
35. Y.-H. Lin, M.-S. Tsai, An advanced home energy management system facilitated by nonintrusive load monitoring with automated multiobjective power scheduling. IEEE Trans. Smart Grid **6**(4), 1839–1853 (2015)
36. Y. Chen, P. Xu, J. Gu, F. Schmidt, W. Li, Measures to improve energy demand flexibility in buildings for demand response (DR): a review. Energy Build. **177**, 125–139 (2018)
37. R.G. Junker, A.G. Azar, R.A. Lopes, K.B. Lindberg, G. Reynders, R. Relan, H. Madsen, Characterizing the energy flexibility of buildings and districts. Appl. Energy **225**, 175–182 (2018)

Chapter 4
Forecasting Available Demand-Side Flexibility

4.1 Overview

The role of flexibility in increasing the efficiency and stability of the grid is an undeniable fact. In flexibility utilization, first step is known to be characterisation, meaning detrmining metrics and indices capable of describing and quantifying flexibility, next step would be forecasting available flexibility. Forecasting Demand-side flexibility refers to the actions which forecast the portion of demand in the system that is changeable or shiftable in response to the signals provided by different entities (e.g., HEMS, aggregator, system operator, etc). In this chapter, first the importance of flexibility then forecasting approaches for the demand-side flexibility are explained.

4.2 Importance of Forecasting Flexibility

Forecasting the flexibility, especially in the demand side, depends on several parameters, including weather forecast (temperature, wind speed, solar radiation), generation, demand, and market prices. There is a clear connection between forecasting the energy generation/consumption and forecasting the weather. Accuracy of generation/demand prediction in the power system and the accuracy of numerical weather prediction models (NWP) depends on each other. In terms of verification and validation, better weather forecast leads to better forecasts for power systems; feedback from the energy sector helps to improve weather prediction models. In this regard, combining all available datasets—in the site, models, satellites, public, private, research is a challenge that must be addressed within big data and data science. More research is needed on understanding how flexibility can be addressed and analyzed via interdisciplinary and multi-agency approaches including energy, weather, data, and artificial intelligence science.

© The Author(s), under exclusive license to Springer Nature Singapore Pte Ltd. 2020   39
R. Ahmadiahangar et al. *Demand-side Flexibility in Smart Grid*,
SpringerBriefs in Applied Sciences and Technology,
https://doi.org/10.1007/978-981-15-4627-3_4

Integrated Energy Systems(IES) bring together a wide range of energy carriers—electricity, thermal sources, and fuels with other infrastructures, such as water, transportation, and data networks. While most energy sources, delivery systems, and demand-response programs are treated as stand-alone technologies today, IES examines how they can optimally work together as a system. By focusing on the optimization of energy systems across multiple pathways and scales, available flexibility can be increased. Therefore, there is a need to strengthen the scientific basis of identifying available flexibility in IES and ways of exploiting it through modifying market frameworks.

4.3 Forecasting Parameters

Ever-increasing power consumption especially in the residential sector increases the importance of creating balance between generation and demand [1]. Unpredicted load consumption and the fluctuation in the power generation of renewable energy sources cause some challenges in this balance [2]. This also causes some challenges in power transmission and distribution [3]. Different strategies and policies have been proposed to address these challenges by the researchers of this field like demand-side management (DSM) and demand-response (DR). Figure 4.1 illustrates the schematic of load management [4]. One of the recent concepts in DR is flexibility that which is provided by distributed energy resources (DER) [5, 6]. Researchers propose different definitions of this concept. Two main definitions of flexibility are as following:

Definition 1 Flexibility can be defined as a response of a DER to a penalty signal considering the usage pattern and power generation [7].

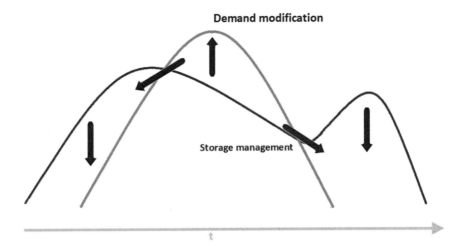

Fig. 4.1 Load and storage management considering the generation and load [10]

Definition 2 This concept in the residential sector is defined as the capacity of changing and adaptation of consumption patterns considering different aspects like the generation and time-varying market price as a penalty signal [8].

Although flexibility can apply in different sectors like commercial, industrial sector and residential sector [9, 10], due to the fact that research shows that more than 30% of usage is for the residential sector, this sector gets a lot of attention in last few years [11]. Increasing the flexibility in this part greatly influences the flexibility of the grid and increases it. Figures 4.2 and 4.3 show the domestic consumption of

Fig. 4.2 UK annual domestic electricity consumption in 2012 [11]

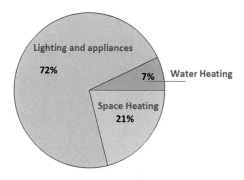

Fig. 4.3 UK domestic electricity consumption—lighting and appliances in 2012 [11]

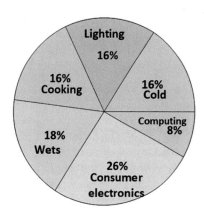

the UK [11]. The studies in this field can be categorized in different classes based on the different assumptions and conditions that each of them has considered:

(1) Different definitions for energy flexibility of a house.
(2) Measuring the flexibility in transition or distribution system operators.
(3) Number of the buildings: Some studies just consider a building and analyze its flexibility. However, others consider the mix of buildings.
(4) Available energy flexibility with residents or without them.

(5) Different appliances of a house are considered to measure energy flexibility like thermal storage, heat pump, batteries, and hot water storage tank. Two main groups of deferrable and thermostatically-controlled loads (TCL) improve the flexibility of the residential sector.
(6) Considering the weather condition, e.g., analyzing separately the energy flexibility of a house during winter and summer.

Lots of strategies have been proposed to increase the flexibility of the grid. Generally, electricity sector can enhance the flexibility considering the following factors:

(i)   The behavior of the residents;
(ii)   The common consumption pattern of the residents and its features;
(iii) The willingness of residents to change the consumption pattern;
(iv)   Measuring the flexibility of each house and monitoring it.

Flexibility can be measured with different algorithms. In the following section, forecasting the flexibility, especially in the demand side, depends on several parameters as following [12, 13]:

(1) The physical characteristics of a building such as type, floor area, year built, U values, thermal mass, insulation, and architectural layout.
(2) The technology of the building such as ventilation.
(3) The control system of the building which is responsive to the penalty signal such as time-varying price.
(4) The weather condition (temperature, wind speed, solar radiation).
(5) The behavior of the occupants and their consumption usage pattern.
(6) The capacity and willingness of the occupants to change their usage pattern.
(7) Comfort level requirement of occupants.
(8) Generation, demand and market prices.

In other words, there is a clear connection between forecasting the energy generation/consumption and forecasting the weather. Accuracy of generation/demand prediction in power system and the accuracy of numerical weather prediction models (NWP) depend on each other. In terms of verification and validation, better weather forecast leads to better forecasts for power systems; feedback from the energy sector helps to improve weather prediction models. In this regard combining all available datasets—in the site, models, satellites, public, private, research is a challenge that must be addressed within big data and data science. More research is needed on understanding how flexibility can be addressed and analyzed via interdisciplinary and multi-agency approaches including energy, weather, data, and artificial intelligence science. All aforementioned parameters should be considered for flexibility measurement. Sometimes, the residents give you these data, but in most cases, these data should be estimated by purely analytical methods. These can be modeled in different ways considering the classic methods or can be achieved by data-driven approaches [14]. In [13] considering some of the aforementioned factors, occupancy matrix is defined as Eq. 4.1 then occupancy or vacancy of a house is extracted based on statistical tools.

$$\text{occupancy matrix} = \begin{pmatrix} \text{household serial number} \\ \text{timeline} \\ \text{sample serial number} \\ \text{sample size} \\ \text{household size} \end{pmatrix}^{\mathrm{T}} \quad (4.1)$$

4.4   Improving Forecasts with Machine Learning Methods

The development of technology and the emergence of different sensors in the smart grid caused the challenge of high volume of data for researchers and engineers of this field. Conventional approaches like computing the average of the data are not efficient in analyzing this volume of data and loss useful information of the data. Therefore, machine learning (ML) approaches are considered as a powerful tool to deal with huge datasets. There are different methods of ML such as classification, clustering, neural network (NN) and so on which were used in this field. As a common method of NN, Recurrent neural networks (RNN), is used in [15] for analyzing the data. This method is based on the feedback of each hidden layer from the previous ones. This algorithm has some initial parameters that should be adjusted by the designer.

Clustering is a common method of segmenting data is used in different studies to compressing big amount of data into few representatives. The flowchart of different methods of clustering is shown in Fig. 4.4 [16]. For analyzing these patterns, authors in [8] calculate and assign a score to each household by using the statistical methods. This score is defined based on three main factors, (I) Consistency of operation, (II) Frequency of operation, and (III) peak-time operation. In other words, this score analyzes the behavior of households and measures their capacity in shifting the load. Hierarchical clustering (linkage-ward clustering algorithm) is used in [17] to compress the huge amount of data in a small volume of data as its representative.

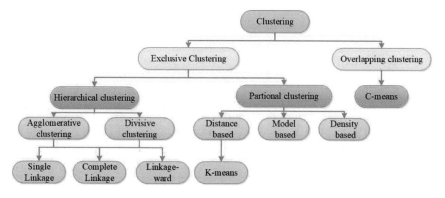

Fig. 4.4   Different methods of clustering [16]

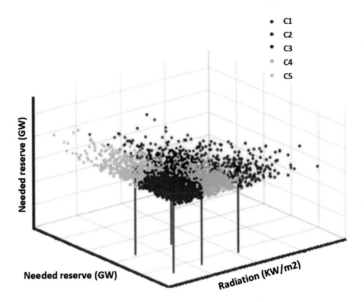

Fig. 4.5 The dataset via its clusters' representatives [18]

This research uses the cluster center aggregation in computing flexibility. In [18] a clustering-based approach is proposed to extract the flexibility of heat pump which its usage pattern is absent. Authors in [19], in order to analyze the effect of uncertainty on the flexibility, generated different scenarios of wind speed, solar radiation, and reserve. For dealing with these scenarios, K-means clustering is applied on 3000 mix dataset with three features, windspeed, solar radiation, and reserve. Figure 4.5 illustrates these datasets via the clusters' representatives. Historical data of a power supply company was clustered in [20] based on k-means algorithm to extract the adjustable loads and flexibility coefficient. Deep embedded clustering is used in [21] to analyze and categorize the usage pattern of consumers to increase the accuracy of flexibility estimation.

4.5 Optimization-Based Approaches

Optimization-based algorithms can be considered as common and useful methods of problem-solving in different fields of smart grid. The total schematic for energy flexibility measurement considering different factors that affect it is displayed in Fig. 4.6 [22]. Authors in [3] considered a power flow problem and calculated the active and reactive flexibility in a specific case study. This proposed power flow based optimization approach has three different inputs, (1) Flexibility inputs (Market-based, DSO own assets, Regulated flexibility), (2) Technical inputs (Current status of grid's equipment, Network topology data, Technical data (e.g., voltage limits, Load,

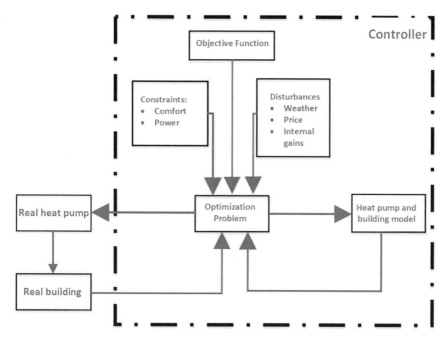

Fig. 4.6 The schematic of optimization-based flexibility computation [22]

and DRES forecasts), and (3) other inputs (Maximum flexibility cost, flexibility activation prices). The output of this optimization is flexibility map and operation point. The effect of price on the active/reactive flexibility area is shown in Fig. 4.7.

MPC-based optimization is proposed in some papers such as [23, 24]. The total schematic of the MPC-based optimization is shown in Fig. 4.8. In [24], first the estimated data of the temperature, load, and generation are updated then demand management plans are implemented. An MPC-based optimization in [25] shows that flexible loads adapt their power usage based on the most economical operation schedule without reducing the comfort of the residents (Fig. 4.8).

4.6 Evaluation Metrics

Different metrics can be used to show the benefits of energy flexibility from different perspectives. Its efficiency can be measured with the ability of load shifting, the economic benefits for the residents and the residents' comfort level. Several researches consider energy flexibility in microgrids and residential buildings and tackle the problem with different approaches, from increasing grid voltage and frequency stability to [26–33]. Authors in [9] considered different evaluation metrics as below:

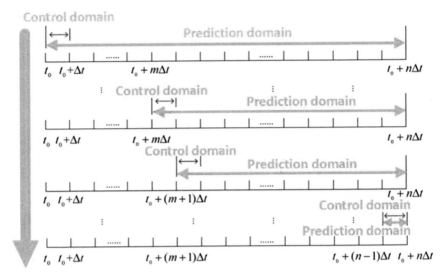

Fig. 4.7 Flexibility cost effect on active/reactive flexibility area [3]

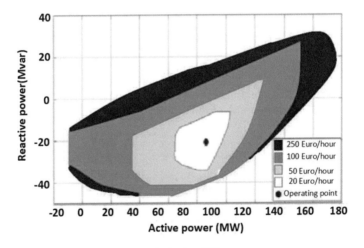

Fig. 4.8 The schematic of MPC-based optimization [23]

(1) Ability of power adjustment

This metric is from the grid's viewpoint which shows the ability to: (1) reduce the power usage during the high-prices hours and contribute to peak shaving at these periods and (2) increase the usage in low-priced hours.

(2) Energy flexibility factor

This metric is based on the role of the thermal mass in the house.

(3) Economic benefit

This factor is from the residents' viewpoint and it is expected that the flexibility gives the residents an opportunity to get economic costs.

(4) Comfort level

Another important factor from the residents' viewpoint is the comfort level. Changing the set-points of the thermal mass, air conditioner or other stuff should not affect the comfort of the residents.

As the conclusion of this chapter, it can be mentioned that in order to utilize energy flexibility in the system, it must be estimated first. For achieving a successful energy flexibility estimation, it is crucial to have accurate forecasts of energy consumptions in different levels (consumer, aggregator) as well as energy generations. Despite recent attention of researchers in this topic, forecasting energy flexibility remains to be a difficult problem. Meanwhile, machine learning based approaches are showing promising results.

Acknowledgements This work has been supported by the European Commission through the H2020 project Finest Twins (grant No. 856602).

References

1. H. Cai et al., Predicting the energy consumption of residential buildings for regional electricity supply-side and demand-side management. IEEE Access **7**, 30386–30397 (2019)
2. M. Ayar et al., A distributed control approach for enhancing smart grid transient stability and resilience. IEEE Trans. Smart Grid **8**(6), 3035–3044 (2017)
3. J. Silva et al., Estimating the active and reactive power flexibility area at the TSO-DSO interface. IEEE Trans. Power Syst. **33**(5), 4741–4750 (2018)
4. P. Kohlhepp et al., Large-scale grid integration of residential thermal energy storages as demand-side flexibility resource: a review of international field studies. Renew. Sustain. Energy Rev. **101**, 527–547 (2019)
5. Zongxiang Lu, Haibo Li, Ying Qiao, Power system flexibility planning and challenges considering high proportion of renewable energy. Autom. Electr. Power Syst. **40**(13), 147–158 (2016)
6. S. Stinner, D. Müller, P. Heiselberg: *Quantifying and aggregating the flexibility of building energy systems*. No. RWTH-2018-224242. E. ON Energy Research Center (2018)
7. A. Wang, R. Li, S. You, Development of a data driven approach to explore the energy flexibility potential of building clusters. Appl. Energy **232**, 89–100 (2018)

8. M. Afzalan, F. Jazizadeh, Residential loads flexibility potential for demand response using energy consumption patterns and user segments. Appl. Energy **254**, 113693 (2019)
9. M. Liu, P. Heiselberg, Energy flexibility of a nearly zero-energy building with weather predictive control on a convective building energy system and evaluated with different metrics. Appl. Energy **233**, 764–775 (2019)
10. N. Ludwig, et al. Industrial demand-side flexibility: a benchmark data set, in *Proceedings of the Tenth ACM International Conference on Future Energy Systems* (2019)
11. Brian Drysdale, Wu Jianzhong, Nick Jenkins, Flexible demand in the GB domestic electricity sector in 2030. Appl. Energy **139**, 281–290 (2015)
12. R.G. Junker et al., Characterizing the energy flexibility of buildings and districts. Appl. Energy **225**, 175–182 (2018)
13. R. Li, et al., in *Energy Flexibility of Building Cluster–Part I: Occupancy Modelling* (2018)
14. Rafał Weron, Electricity price forecasting: A review of the state-of-the-art with a look into the future. Int. J. Forecast. **30**(4), 1030–1081 (2014)
15. Kaveh Paridari, Lars Nordström, Flexibility prediction, scheduling and control of aggregated TCLs. Electr. Power Syst. Res. **178**, 106004 (2020)
16. E. Azizi et al., Application of comparative strainer clustering as a novel method of high volume of data clustering to optimal power flow problem. Int. J. Electr. Power Energy Syst. **113**, 362–371 (2019)
17. D. Patteeuw et al., Clustering a building stock towards representative buildings in the context of air-conditioning electricity demand flexibility. J. Build. Perform. Simul. **12**(1), 56–67 (2019)
18. K. Kouzelis et al., Estimation of residential heat pump consumption for flexibility market applications. IEEE Trans. Smart Grid **6**(4), 1852–1864 (2015)
19. A. Alirezazadeh et al., A new flexible model for generation scheduling in a smart grid. Energy **191**, 116438 (2020)
20. S. RongQi, et al., Research of flexible load analysis of distribution network based on big data, *2019 IEEE 4th International Conference on Cloud Computing and Big Data Analysis (ICCCBDA)* (IEEE, 2019)
21. M. Sun et al., Clustering-based residential baseline estimation: a probabilistic perspective. IEEE Trans. Smart Grid **10**(6), 6014–6028 (2019)
22. T.Q. Péan, S. Jaume, R. Costa-Castelló, Review of control strategies for improving the energy flexibility provided by heat pump systems in buildings. J. Process Control **74**, 35–49 (2019)
23. C. Lv et al., Model predictive control based robust scheduling of community integrated energy system with operational flexibility. Appl. Energy **243**, 250–265 (2019)
24. T. Péan, J. Salom, R. Costa-Castelló, Configurations of model predictive control to exploit energy flexibility in building thermal loads, in *2018 IEEE Conference on Decision and Control (CDC)* (IEEE, 2018)
25. G. Chen, D. Liu, Y. Lixia, Predictive control of regional flexible load cluster based on mixed logical dynamic method, in *2019 IEEE Innovative Smart Grid Technologies-Asia (ISGT Asia)* (IEEE, 2019)
26. R. Ahmadiahangar, A. Rosin, A. NabaviNiaki, I. Palu, T. Korõtko, A review on real-time simulation and analysis methods of microgrids. Int. Trans. Electr. Energy Syst. **29**(11), e12106 (2019)
27. R. Ahmadiahangar, T. Häring, A. Rosin, T. Korõtko, J. Martins, Residential load forecasting for flexibility prediction using machine learning-based regression model, in *2019 IEEE International Conference on Environment and Electrical Engineering and 2019 IEEE Industrial and Commercial Power Systems Europe (EEEIC/I&CPS Europe)*, 11 June 2019, pp. 1–4 (IEEE)
28. K. Peterson, R. Ahmadiahangar, N. Shabbir, T. Vinnal, Analysis of microgrid configuration effects on energy efficiency, in *2019 IEEE 60th International Scientific Conference on Power and Electrical Engineering of Riga Technical University (RTUCON)*, 7 Oct 2019, pp. 1–6 (IEEE)
29. M. Mahmudizad, R.A. Ahangar, Improving load frequency control of multi-area power system by considering uncertainty by using optimized type 2 fuzzy pid controller with the harmony search algorithm. World Acad. Sci. Eng. Technol. Int. J. Electr. Comput. Energ. Electron. Commun. Eng. **10**(8), 1051–1061 (2016)

30. T. Häring, R. Ahmadiahangar, A. Rosin, H. Biechl, Impact of load matching algorithms on the battery capacity with different household occupancies, in *IECON 2019-45th Annual Conference of the IEEE Industrial Electronics Society* (Lisbon, 2019)
31. N. Shabbir, R. Ahmadiahangar, L. Kütt, A. Rosin, Comparison of machine learning based methods for residential load forecasting, in *2019 Electric Power Quality and Supply Reliability Conference (PQ) & 2019 Symposium on Electrical Engineering and Mechatronics (SEEM)*, 12 June 2019, pp. 1–4 (IEEE)
32. R. Ahmadi, A. Sheikholeslami, A. Nabavi Niaki, A. Ranjbar, Dynamic participation of doubly fed induction generators in multi-control area load frequency control. Int. Trans. Electr. Energy Syst. **25**(7), 1130–1147 (2015)
33. R. Ahmadi, F. Ghardashi, D. Kabiri, A. Sheykholeslami, H. Haeri, Voltage and frequency control in smart distribution systems in presence of DER using flywheel energy storage system, *IET Digital Library* (2013)

Chapter 5
New Approaches for Increasing Demand-Side Flexibility

5.1 Overview

In the case of conventional power plants, increasing the flexibility can increase generation cycling, which can influence air, water, and solid waste emissions. In hydropower plants, increasing flexibility raise concerns about creating serious ecological impacts on a local scale. However, hydroelectric production has increased to at least 20% of energy consumption coming from RES following the demand trend and EU directives' targets for 2020 [1, 2].

Determining environmental impacts associated with the transition to a more flexible power system is challenging due to the diverse scenarios of generation, delivery, storage, demand management, and energy efficiency measures and it needs collaboration of researchers from multidisciplinary fields with industries and stack holders.

Therefore, increasing the demand-side flexibility, has been a recent topic that attracts much attention, this chapter discuss new approaches for increasing the demand-side flexibility. It worth mentioning that increasing the demand-side flexibility strongly relies on the social acceptance of customers. Social acceptance itself depends on energy prices and the way it affects the customer. However, there is a great connection between social acceptance, market prices, and demand response programs.

© The Author(s), under exclusive license to Springer Nature Singapore Pte Ltd. 2020
R. Ahmadiahangar et al. *Demand-side Flexibility in Smart Grid*,
SpringerBriefs in Applied Sciences and Technology,
https://doi.org/10.1007/978-981-15-4627-3_5

5.2 Enabling Demand-Side Flexibility

Demand-side flexibility is a new topic that has not been addressed much. Most researches consider the possibility of increasing PS flexibility from different generation technologies. The study in [3] presents a techno-economic analysis of flexibility from distributed energy resources and the economic valuation of flexibility for trading in traditional markets considering a case from the Netherlands. This paper presents flexibility costs calculation methods from distributed energy resources and concluded that demand management and flywheel technologies are the most economical options considering current market situations.

Demand has a significant potential to contribute to the flexibility of the PS, from quickly responding to supply shortages, to following price signals to change the demand profile so that energy is consumed when it is cheaper to supply and when the grid does not face congestion [4]. Flexibility in demand-side is the ability of consumption modification in response to control signals [5]. Possible sources of those control signals may be external market signals to the smart meters or internal control signals from the home energy management system (HEMS). Typically, residential consumers account for 40% of total consumption, therefore, there is a significant potential to utilize flexibility in the demand-side [6].

Demand response (DR) programs, like direct load control, time-of-use (TOU), and real-time pricing are used to meet the energy demand and available power, thus improving grid stability to mitigate the adverse effects of high price volatility for both the utility and the consumers. Some researches proposed using artificial intelligence in DR management programs to increase flexibility exploitation [7]. The study in [8] estimates demand-side flexibility by applying thermostat setpoint changes in commercial and residential buildings. The study in [9] investigates heat pump control strategies for enhancing energy flexibility in buildings. However, there is a trend toward exploiting flexibility from newer kinds of demands like electrical vehicles [10, 11]. Potentials of using installed distributed energy resources in the demand-side for improving flexibility are discussed in [12]. In applying DR programs, another ongoing challenge is how to maintain or even improve customer's comfort level, while maximizing profit of the program [13].

Customers are proposed to potentially being able to control their consumption profile according to grid operational conditions in [14]. The study in [15] investigates the potential technologies and solutions like energy storage systems and demand response programs that can provide flexibility in Finnish PS. The flexibility potential of the smart appliances, or the maximal amount of time a specific increase or decrease of power is realized within the comfort requirements of 186 households, during 3 years in Belgium, is calculated in [16]. The study in [17] shows that even with 5% smart meter coverage, one can forecast, with high confidence, the composition of the load at the substation, results are based on research in the UK grid. The system operator in the UK is planning to obtain 30–50% of balancing services through DR [18]. In the US, it was estimated that residential customers' participation in DR might bring up to half of the total peak reduction [19].

Managing DR programs to PS stability and cost-effectiveness require advanced data analytics for acquiring accurate information and automated decision support and handling events in a timely fashion. Recently, more progress have been made for utilizing field data extracted from smart metering devises in substations, feeders, and various databases and models across the utility enterprises and smart meters installed in individual customers. Large amounts of data are increasingly accumulated in the energy sector with the continuous application of sensors, wireless transmission, network communication, and cloud computing technologies. Data analysis is one of the critical factors that will determine the success of smart meters, which deals with data acquisition, transmission, and processing [20]. To fulfill the potential of energy big data and obtain insights to achieve smart energy management, comprehensive studies of big data-driven smart energy management were investigated [21]. Also, through business intelligence and analytics paradigm, the concept of cost analysis and pricing could be helpful [22]. The development of a novel data-driven Demand-Side Management (DSM), whose framework includes demand forecasting [23, 24] customer response analysis, and prediction of dynamic condition of the energy network, quick supply reliability evaluation, multi-objective optimization, and decision-making is presented in [25]. Moreover, emerging blockchain technologies are presented to provide the seamless and secure implementation of a decentralized demand-side management approach to be implemented in practice to facilitate peer-to-peer transactions following optimized demand profiles [26].

The study in [27] characterizes the underlying sequential decision-making problem as a Markov decision process and uses Reinforcement Learning techniques to solve it. Capturing the underlying hidden features of thermostatically controlled loads and extracting their flexibility is proposed in [28]. Novel load management scheme for highly stochastic loads for DSM is also developed based on the Markovian method in which each household possesses individually suited parameters [29].

From the management and control side of view, recently, controllability of PS has been significantly improved by the increasing integration of energy storage and power electronic devices in the grid. However, mainly due to the complicated coordination of these various controllable resources, the controllability cannot be fully translated into the system flexibility.

Currently, in practice, the demand response solutions implemented in a significant number of countries do not consider aggregation of customers/prosumers at Low Voltage (LV) level but typically focus on fewer resources of greater individual size (i.e., industrial loads) connected to Medium and High Voltage levels [30]. The integration of demand flexibility in distributed generation (DG) planning either lacks accuracy or ignores the potential of the behavior of consumers in promoting the integration of renewables. In the US, 29.5 GW of DSM is already available to market participants. The situation in Europe is promising, but yet in an early stage of development. Countries like Belgium, Great Britain, Denmark, Finland, France, Ireland, and Switzerland are already in a level of commercial Demand Response (DR) product offerings. The rest of Europe states are still bounded to national regulations, which make DSM services infeasible. Activating the DR flexibility of prosumers' households is a major step towards better RES utilization and more efficient system operation [31]. Nevertheless, activating flexibility in individual prosumers is not effective enough without finding proper solutions on how to trade it the energy markets. Aggregator, as a new market entity, which is based on Virtual Power Plant (VPP), is the answer to this question [2].

In order to enable active participation of demands, the system requires a new actor to manage the resources connected at LV level in the most efficient way [32]. Demand-side flexibility of thermal systems has been considered to provide frequency regulation services in a hierarchical market with a bi-level optimization approach [33]. A hierarchical control framework is proposed in [34] that is characterized by decentralized decision-making, can tackle the privacy concerns of consumers and is capable of aggregating flexibility in the wholesale market. The study in [35] proposes a market-based framework to manage multiple flexibility services. A Decentralized Markets design is proposed in [36] distribution system operator to manage local demand constraints by obtaining flexibility from competing aggregators, which must in turn incentivize prosumers to provide this flexibility.

Some recent European Union projects including EMPOWER H2020 [37] and INVADE H2020 [38] aims to enable prosumers to participate in different market services. The study in [39] quantifies the net revenues that can be captured by a flexible resource able to react to the short-term price variations on the day-ahead and intraday markets in Germany. The study in [40] in the UK shows that the expansion capacity of renewable generation can be significantly increased with a new DG planning model coordinating demand flexibility.

Penalty-based control is also an option to enable the flexibility from demand-side without violating flexibility concerns [41]. Utilizing individual penalty signal is the proposed solution in [41] to increase the flexibility exploitation.

In a real-time electricity market, the market-clearing price is determined by the random deviation of actual power supply and demand from the predicted values in the day-ahead market [42]. Thus, dynamically increasing the availability of flexibility from one side will reduce the volatility of the price and from the other side increases the profit of demand response programs.

The main disadvantage of typical analyzing tools of PS (software simulations, prototypes and pilot projects) is the limited ability to solve technical and economic problems at once; hence, several simplifications must be taken. In this context, Real-Time (RT) simulations and Hardware-in-the-loop (HIL) technology are beneficial mainly because of their easily reconfigurable test environment [43]. In latter, all system variables are accessible and there is a good possibility of testing different scenarios, including and cases with the same hardware setup [44, 45]. It is also worth mentioning that a real-time simulation is a promising approach for validating advance and complex control strategies designed for PS and also determining exact values of control parameters and debugging them. As a result, significant cost reduction can be achieved by eliminating control system errors [46]. Particularly, for flexibility studies, it is always needed to consider models from different domains, also analysis and related metrics are supposed to be real-time. Therefore, RT simulations are probably the best options for design and validating of control systems to deliver flexibility in demand-side.

5.3 Electrical Vehicle (EV) Charging as Alternative Storage [47]

An average EV battery capacity could sustain one regular household's daily electricity demand. However, as the majority of recent commercially available EVs are unable to generate electricity towards the grid, their controlled charging would be able to provide only valuable storage capacity or balancing load when there is significant overproduction in the system. Therefore, HEMS can use this capacity as a source of flexibility. In ideal situation, this flexibility can be utilised through the aggregator.

According to IEA [48], in Finland, approximately 60 percent of the cars have an engine block heater; which could be used for EV charging as well as for integration into the power system as energy storages with slight change. This way most of the EV owners have access to a standard charger, which would serve for charging plug-in vehicles [49].

possible charging impact on the load has been divided into three levels:

Standard charging considers the market price fluctuations and traffic intensity. Equation (5.1) is used to calculate the hourly load impact or also the storage capacity (MWh/h) of standard charging outside the prime time charging period [47]:

$$E_{C_h} \atop h=7,...,19 = \frac{\sum_{l=20}^{6} d \cdot \rho_{T_h} \cdot E_d}{\sum_{i=7}^{19} 1_i} \cdot \Delta V_{P_h} \qquad (5.1)$$

where h is the hour of the day; l defines the prime time hours; i is the time period outside the prime charging time; E_{C_h} is the charging energy of the hour; d is the total daily distance covered by EVs (km); ρ_{T_h} is hourly traffic density; E_d is the energy consumption per distance covered ($2.35 \cdot 10^{-5}$ MWh/km), and ΔV_{P_h} is the relative change in the number of parking EVs to the average parking EVs over the observed hours of the day (%/h). The values of h, l, and i used in the equation should match the prime time selection.

Equation (5.2) is used to calculate the hourly impact of standard charging during the prime time [47].

$$E_{C_h} \atop h=20,...,6 = a \cdot d \cdot E_d \cdot D(MP)_h \qquad (5.2)$$

where h is the hour of the day; E_{C_h} is the charging energy of the hour; a is the share of EV owners (%) who have access to a standard charging station (block heater); d is the total daily distance covered by EVs; E_d is the energy consumption per distance covered, and $D(MP)_h$ is the charging distribution coefficient during the prime charging time (%/h). The values of h used in the equation should match the prime time selection.

Comfort charging is the energy used for preheating the car and the battery two hours prior to usage in order to maximize the battery usage for driving and not to heat the cabin in cold battery condition. The preheating is assumed to have an energy intensity of 3×10^{-3} MWh per car in one hour and could serve as an extra storage possibility during autumn and winter periods (e.g. the Nissan Leaf's heater power remains between 1.5 and 3 kW). Equation (5.3) describes how to calculate the hourly energy required for preheating [47]:

$$E_{P_h} = (\Delta V_{(h+2)} \cdot E_H)\big|_{h<(h_n-1)} + (\Delta V_{(h+1)} \cdot E_H)\big|_{h>(h_{m-2})} \qquad (5.3)$$

where h is the hour of the day; E_{P_h} is the preheating energy of the hour; ΔV_h is the change of vehicles in traffic during an observed hour (1/h); E_H is the hourly consumption of the preheaters; h_m is the hour when the increase of traffic intensity starts, and h_n is the last hour of the cycle when the traffic intensity has increased.

Quick charging takes into account the car users who do not have full-time access to standard chargers and thus need to utilize the quick charging infrastructure. This load is calculated with (5.4):

$$E_{Q_h} = (d \cdot E_d - E_C) \cdot \rho_{T_h} \qquad (5.4)$$
$$h=0,...,23$$

where h is the hour of the day; E_{Q_h} is the charging energy of the hour; d is the total daily distance covered by EVs; E_C is the total energy-charged via the standard chargers, and ρ_{T_h} is the hourly traffic density (%/h).

According to the proposed Eqs. (5.1)–(5.4) EV fleet consisting of 100,000 cars during a winter day consumption would remain around 1113 MWh. The hourly distribution of the load over a 24 h period is shown in Fig. 5.1.

5.4 Increasing Available Flexibility; Case Study, Estonia

Increasing available flexibility through multi-energy communities or integrated energy systems is also proposed in some researches [50], however, a new trend is to maximize exploitation of existing flexibility on PSs [51–53].

In 2019, generation of electricity from renewable sources grew by 17 per cent year-over-year to 1.9 terawatt-hours in Estonia, while non-renewable production fell by nearly one-half, to 4.5 terawatt-hours. Renewable energy made up 21 per cent of total consumption. As the share of variable renewables rises, so does operational complexity. In order to reach the European Commission's targets for 2050, the integration of renewable Energies will require flexibility sources, independent of conventional generation, in order to provide standard security of supply.

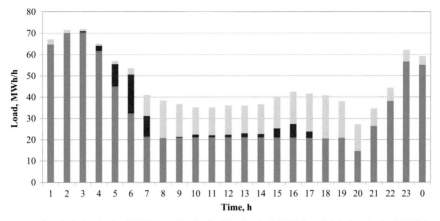

Standard charging load, MWh/h ■ Comfort load (preheating), MWh/h ■ Quick charging load, MWh/h

Fig. 5.1 Hourly distribution of the load required by an EV fleet of 100,000 according to different load sources [47]

Oil shale power plants have been the main provider of generation in Estonia with a generation share of almost 90%. In recent years share of renewable energy sources is increased to 30%, mostly coming from the wind turbines and biomass-burning cogeneration power plants [54]. Estonia's strongest potential in renewable energy lies in bioenergy-based combined heat and power generation, in wind power and the production of biomethane, which possesses qualities identical to natural gas and as such can be used as a replacement for natural gas. Small-scale hydro and solar power capacity is also being developed [55].

The annual electricity consumption of Estonia is about 8 TWH. To this point, overall, power consumption is showing a growth trend, but the peak loads on the power grid have remained essentially unchanged in the last decade—between 1500 and 1710 MW. The total installed net generating capacity is 2828 MW and of the capacity usable during peak periods 1848 MW. Estonian grid is interconnected to Finland through total of 1000 MW of DC connections, to Latvia with 800 MW of AC connections. There are also three 330 kV interconnections with Russia that are just utilized for balancing power and frequency regulations [55].

Electrical heating is playing a decreasing role in the overall electricity consumption in private households. The largest DSO, Elektrilevi, has done an estimation by observing demand changes in different outside temperature situations, which puts the amount of electric power used for heating purposes around 800 GWh. The oil shale power plants along with the interconnection capacities have so far been adequate to provide flexibility to the system; however, their role in the Estonian power system can be seen decreasing in the future [56].

Although condensing oil shale power plants will hardly be fully phased out in the near future, it is necessary to look at new sources of flexibility to compliment biomass cogeneration power plants and the increasing share of wind power. Furthermore, the need for flexibility services in the Baltic countries may increase drastically when they are decoupled from the IPS/UPS system [57].

Estonian two-month wind generation and forecast are shown in Fig. 5.2. As the share of variable renewables rises, so does operational complexity. In order to reach the European Commission's targets for 2050, the integration of renewable Energies will require flexibility sources, independent of conventional generation, in order to provide standard security of supply. Figure 5.3 shows Estonian two-month power consumption/production [55].

Price volatility, swings between low and high prices, is the main sign of lack of flexibility in the grid. Price volatility can reflect limited grid infrastructure capacity,

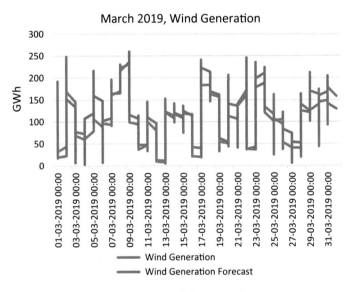

Fig. 5.2 Estonian two-month wind generation and forecast [55]

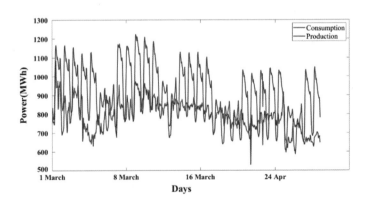

Fig. 5.3 Estonian two-month power consumption/production [55]

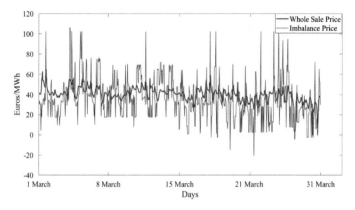

Fig. 5.4 Estonian one-month day-ahead wholesale and imbalance price [55]

limited availability of ramping, fast response, and peaking supplies, and limited demand response. Analyzing Estonian market balancing prices in 2018–2019, we noticed a pattern of volatility and negative prices. Figure 5.4 shows a sample one-month price with signs of negative prices and volatility in imbalance price and the day-ahead market price for the same period (March 2019 [55]).

5.5 Challenges in Increasing Available Flexibility

One challenge, particularly for the European electricity market is low prices—subsidies and other administrative measures lower the price of electricity artificially. Low price sensitivity is another challenge, since it makes flexibility utilisation unattractive for the conconsumers. Flexibility price making in different markets, e.g., day-ahead, futures, intraday, balancing energy can be improved so the market can find the right balance [55]. It should be also pointed out that, increasing demand-side flexibility strongly relies on social acceptance of customers. Social acceptance itself depends on energy price and the way it affects customer. With enabling flexibility trade between customers and the aggregator, decreasing customers' electricity costs, it would be possible to increase social acceptance of such programs.

As the conclusion, characterisation, forecast and control of energy flexibility in demand-side are key issues to be solved. Flexibility characterisation describes the shape of flexibility and provides proper metrics for evaluation. Determining metrics and characterisation methods for demand-side flexibility is an ongoing hot topic for research. Next step is to forecast available flexibility in demand side, meaning, and determining amount of flexibility in specific time steps. Machine learning based methods are promising approaches for forecasting demand-side flexibility. Finally, novel control systems and market frameworks are needed to deliver forecasted flexibility in the system.

Acknowledgements This work has been supported by the European Commission through the H2020 project Finest Twins (grant No. 856602).

References

1. R. Ahmadiahangar, A. Sheikholeslami, A. Nabavi Niaki, A. Ranjbar, Dynamic participation of doubly fed induction generators in multi-control area load frequency control. Int. Trans. Electr. Energy Syst. **25**(7), 1130–1147 (2015)
2. M. Mahmudizad, R. Ahmadiahangar, Improving load frequency control of multi-area power system by considering uncertainty by using optimized type 2 fuzzy pid controller with the harmony search algorithm. World Acad. Sci. Eng. Technol. Int. J. Electr. Comput. Energ. Electron. Commun. Eng. **10**(8), 1051–1061 (2016)
3. Cherrelle Eid, Joep Grosveld, Rudi Hakvoort, Assessing the costs of electric flexibility from distributed energy resources: a case from the Netherlands. Sustain. Energy Technol. Assessments **31**, 1–8 (2019)
4. N. Good, K.A. Ellis, P. Mancarella, Review and classification of barriers and enablers of demand response in the smart grid. Renew. Sustain. Energy Rev. **72**, 57–72 (2017)
5. *Regulatory Recommendations for the Deployment of Flexibility*. SGTF-EG3 Report (2015)
6. K. Kouzelis, Z.H. Tan, B. Bak-Jensen, J.R. Pillai, E. Ritchie, Estimation of residential heat pump consumption for flexibility market applications. IEEE Trans. Smart Grid **6**(4), 1852–1864 (2015)
7. P.U. Herath, V. Fusco, M.N. Cáceres, G.K. Venayagamoorthy, S. Squartini, F. Piazza, J.M. Corchado, Computational intelligence-based demand response management in a microgrid. IEEE Trans. Ind. Appl. **55**(1), 732–740 (2018)
8. Ke Wang, Rongxin Yin, Liangzhong Yao, Jianguo Yao, Taiyou Yong, A two-layer framework for quantifying demand response flexibility at bulk supply points. IEEE Trans. Smart Grid **9**(4), 3616–3628 (2018)
9. T. Péan, R. Costa-Castelló, E. Fuentes, Experimental testing of variable speed heat pump control strategies for enhancing energy flexibility in buildings. IEEE Access (2019)
10. I. Pavić, T. Capuder, I. Kuzle, A comprehensive approach for maximizing flexibility benefits of electric vehicles. IEEE Syst. J. **12**(3), 2882–2893 (2018)
11. A.S. Gazafroudi, J.M. Corchado, A. Keane, A. Soroudi, Decentralised flexibility management for EVs. IET Renew. Power Gener. **13**(6), 952–960 (2018)
12. I.I. Avramidis, V.A. Evangelopoulos, P.S. Georgilakis, Demand side flexibility schemes for facilitating the high penetration of residential distributed energy resources. IET Gener. Transm. Distrib. **12**(18), 4079–4088 (2018)
13. A. Taşcıkaraoğlu, N.G. Paterakis, O. Erdinç, J.P. Catalao, Combining the flexibility from shared energy storage systems and DLC-based demand response of HVAC units for distribution system operation enhancement. IEEE Trans. Sustain. Energy **10**(1), 137–148 (2018)
14. M. Inês Verdelho, R. Prata, D. Koraki, Demand flexibility benefits from the DSO perspective—A sustainable case study, in *CIRED Workshop—Helsinki*, 14–15 June 2016
15. Satu Paiho, Heidi Saastamoinen, Elina Hakkaraine, Increasing flexibility of Finnish energy systems—a review of potential technologies and means. Sustain. Cities Soc. **43**, 509–523 (2018)
16. R. D'hulst, W. Labeeuw, B. Beusen, S. Claessens, G. Deconinck, K. Vanthournout, Demand response flexibility and flexibility potential of residential smart appliances: experiences from large pilot test in Belgium. Appl. Energy **155**, 79–90 (2015)
17. Jelena Ponocko, Jovica V. Milanovic, Forecasting demand flexibility of aggregated residential load using smart meter data. IEEE Trans. Power Syst. **33**(5), 5446–5455 (2018)

18. C. C. C4.112, *Guidelines for Power Quality Monitoring—Measurement Locations, Processing and Presentation of Data* (London, UK, 2014)
19. M. Pipattanasomporn, M. Kuzlu, S. Rahman, Y. Teklu, Load profiles of selected major household appliances and their demand response opportunities. IEEE Trans. Smart Grids **5**(2), 742–750 (2014)
20. Smart electricity meter data intelligence for future energy systems: a survey. IEEE Trans. Industr. Inf. **12**(1), 425–437 (2016)
21. K. Zhou, C. Fu, S. Yang, Big data driven smart energy management: from big data to big insights. Renew. Sustain. Energy Rev. **56**, 215–225 (2016)
22. E. Bagheri, H. Fazlollahtabar, M. Talebi Ashoori, Product pricing with marketing data under risk using business intelligence. Revista Inteligência Competitiva **8**(3), 1–14 (2018)
23. N. Shabbir, R. Ahmadiahangar, L. Kütt, M.N. Iqbal, A. Rosin, Forecasting short term wind energy generation using machine learning, in *2019 IEEE 60th International Scientific Conference on Power and Electrical Engineering of Riga Technical University (RTUCON)* (2019)
24. A.B. Dayani, H. Fazlollahtabar, R. Ahmadiahangar, A. Rosin, M.S. Naderi, M. Bagheri, Applying reinforcement learning method for real-time energy management, in *2019 IEEE International Conference on Environment and Electrical Engineering and 2019 IEEE Industrial and Commercial Power Systems Europe (EEEIC/I&CPS Europe)* (2019)
25. H. Su, E. Zio, J. Zhang, L. Chi, X. Li, Z.A. Zhang, Systematic data-driven demand side management method for smart natural gas supply systems. Energy Convers. Manag. **185**, 368–383 (2019)
26. S. Noor, W. Yang, M. Guo, K.H. van Dam, X. Wang, Energy Demand Side Management within micro-grid networks enhanced by blockchain. Appl. Energy **228**, 1385–1398 (2018)
27. F. Ruelens, B.J. Claessens, S. Quaiyum, B. De Schutter, R. Babuška, R. Belmans, Reinforcement learning applied to an electric water heater: from theory to practice. IEEE Trans. Smart Grid **9**(4), 3792–3800 (2018)
28. B.J. Claessens, P. Vrancx, F. Ruelens, Convolutional neural networks for automatic state-time feature extraction in reinforcement learning applied to residential load control. IEEE Trans. Smart Grid **9**(4), 3259–3271 (2018)
29. E. Thomas, R. Sharma, Y. Nazarathy, Towards demand side management control using household specific Markovian models. Automatica **101**, 450–457 (2019)
30. P. Olivella-Rosell, E. Bullich-Massagué, M. Aragüés-Peñalba, A. Sumper, S.Ø. Ottesen, Optimization problem for meeting distribution system operator requests in local flexibility markets with distributed energy resources. Appl. Energy **210**, 881–895 (2018)
31. T. Häring, R. Ahmadiahangar, A. Rosin, H. Biechl, T. Korõtko, Comparison of the impact of different household occupancies on load matching algorithms, in *2019 Electric Power Quality and Supply Reliability Conference (PQ) & 2019 Symposium on Electrical Engineering and Mechatronics (SEEM)* (2019)
32. Gianluca Lipari, Gerard Del Rosario, Cristina Corchero, Ferdinanda Ponci, A real-time commercial aggregator for distributed energy resources flexibility management. Sustain. Energy Grids Netw. **15**, 63–75 (2018)
33. Sareh Agheb, Xiaoqi Tan, Bo Sun, Contract design for aggregating, trading, and distributing reserves in demand-side frequency regulation. IEEE Trans. Industr. Inf. **14**(6), 2539–2550 (2018)
34. I. Lampropoulos, T. Alskaif, J. Blom, W. van Sark, A framework for the provision of flexibility services at the transmission and distribution levels through aggregator companies. Sustain. Energy Grids Netw. **17**, 100187 (2019)
35. P. Olivella-Rosell, P. Lloret-Gallego, Í. Munné-Collado, Local flexibility market design for aggregators providing multiple flexibility services at distribution network level. Energies **11**, 822 (2018)
36. Thomas Morstyn, Alexander Teytelboym, Malcolm D. McCulloch, Designing decentralized markets for distribution system flexibility. IEEE Trans. Power Syst. **34**(3), 2128–2140 (2019)

37. C. Cordobés, *EMPOWER H2020 Project Grant Agreement 646476; eSmart Systems* (Bellevue, WA, USA, 2017)
38. I. Ilieva, B. Bremdal, S. Ødegaard Ottesen, J. Rajasekharan, P. Olivella-Rosell, Design characteristics of a smart grid dominated local market, in *Proceedings of the CIRED Workshop*, Helsinki, Finland, pp 1–4, 14–15 June 2016
39. S. Goutte, P. Vassilopoulos, The value of flexibility in power markets. Energy Policy **125**, 347–357 (2019)
40. C. Dang, X. Wang, X. Wang, F. Li, B. Zhou, DG planning incorporating demand flexibility to promote renewable integration. IET Gener. Transm. Distrib. **12**(20), 4419–4425 (2018)
41. Designing individual penalty signals for improved energy flexibility utilisation. Int. Fed. Autom. Control 52(4), 123–128 (2019)
42. Demand response management for profit maximizing energy loads in real-time electricity market. IEEE Trans. Power Syst. 33(6), 6387–6396 (2019)
43. K. Peterson, R. Ahmadiahangar, N. Shabbir, T. Vinnal, Analysis of microgrid configuration effects on energy efficiency, in *2019 IEEE 60th International Scientific Conference on Power and Electrical Engineering of Riga Technical University (RTUCON)* (2019)
44. M. Caserza Magro, M. Giannettoni, P. Pinceti, Real time simulator for microgrids. Electr. Power Syst. Res. **160**, 381–396 (2018)
45. F. Huerta, R.L. Tello, M. Prodanovic, Real-time power-hardware-in-the-loop implementation of variable-speed wind turbines. IEEE Trans. Industr. Electron. **64**(3), 1893–1904 (2016)
46. R. Ahmadiahangar, A. Rosin, A.N. Niaki, I. Palu, T. Korõtko, A review on real-time simulation and analysis methods of microgrids. Int. Trans. Electr. Energy Syst. **29**(11), e12106 (2019)
47. A. Rosin, I. Drovtar, J. Kilter, Solutions and active measures for wind power integration, in *Large Scale Grid Integration of Renewable Energy Sources* (IET, 2017)
48. International Energy Agency. Hybrid & Electric Vehicle Implementing Agreement [Online]. Available from http://www.ieahev.org/. Accessed 24 Aug 2016 [Online]
49. H. Turker, S. Bacha, D. Chatroux, Impact of plug-in hybrid electric vehicles (PHEVs) on the french electric grid, in *2010 IEEE PES Innovative Smart Grid Technologies Conference Europe (ISGT Europe)* (Gothenburg, 2010)
50. N. Good, P. Mancarella, Flexibility in multi-energy communities with electrical and thermal storage: a stochastic, robust approach for multi-service demand response. IEEE Trans. Smart Grid **10**(1), 503–514 (2019)
51. R. Ahmadiahangar, T. Häring, A. Rosin, T. Korõtko, J. Martins, Residential load forecasting for flexibility prediction using machine learning-based regression model, in *2019 IEEE International Conference on Environment and Electrical Engineering and 2019 IEEE Industrial and Commercial Power Systems Europe (EEEIC/I&CPS Europe)*, Genoa, Italy (2019)
52. N. Shabbir, R. Ahmadiahangar, L. Kütt, A. Rosin, Comparison of machine learning based methods for residential load forecasting, in *2019 Electric Power Quality and Supply Reliability Conference (PQ) & 2019 Symposium on Electrical Engineering and Mechatronics (SEEM)* (2019)
53. T. Häring, R. Ahmadiahangar, A. Rosin, H. Biechl, Impact of load matching algorithms on the battery capacity with different household occupancies, in *IECON 2019-45th Annual Conference of the IEEE Industrial Electronics Society* (Lisbon, 2019)
54. Statistics Estonia, 2019. [Online]. Available: www.stat.ee. Accessed 2019
55. "Elering," 2019. [Online]. Available: www.elering.ee. [Accessed 2019]
56. L. Söder, P.D. Lund, H. Koduvere, T.F. Bolkesjø, A review of demand side flexibility potential in Northern Europe. Renew. Sustain. Energy Rev. **91**, 654–664 (2018)
57. G. Power, *Feasibility Study on the Inter-Connection Variants for the Integration of the Baltic States to the EU Internal Electricity Market Gothia Power* (2015)
58. https://www.elering.ee," [Online]. Accessed Apr 2019

Printed in the United States
By Bookmasters